GENERAL MATH OF NATURE PHENOMENA

SAM GUELLER

About the Software: To install the software, make a directory with name Polyn, insert the diskette in a drive and copy Infer.Exe and Polyn.Exe in this directory. The program Infer is for use with words and Polyn is a program to be used with numbers. Once an output file was created, import it using a wordprocessor, spreadsheets or grapher for further analysis. All graphs were printed with AXUM from MathSoft, Inc., with other graphers results can be seen differently. For words or statements proceed as follows: type Infer and press Enter to run the program. You have the option to input data up to ten words or statements. Bounds are the maximum permutations, for 3 words it is 1 x 2 x 3 = 6 and the bounds are 1 and 6. In this case enter the data when calling for data 3, data 2, and data 1. The program can run directly Microsoft Word or WordPerfect if you have any of these wordprocessor in your computer. For numbers proceed as follows: type Polyn and press Enter to run the program. You have the option to input data up to eight numbers, and eight constants or ranges for the coefficients of the polynomials. The procedure is the same as above for words. For more than eight data numbers use the method described in Numbers as Data. With the option Ranges, the output can exceed the memory capacity of your computer, use Ranges that your computer can handle. In Windows, you can launch both programs with Start-Run, or insert an icon in the Windows front page with My Computer – Explore – click the file *.exe – drag to the front page, and "Create an Icon Here". The program can run directly the Axum grapher if you have it in your computer. Figures i through iii are monitor views of both programs.

ISBN – 13:978-1540793720

ISBN-10-1540793729

TABLE OF CONTENTS

INTRODUCTION

As occurrence in a phenomenon of nature follows a Normal Distribution, it has been proven they also can itself be described by means of a Polynomial, therefore a polynomial can be used always as a law of a nature phenomenon. Both, the Normal Distribution and Polynomials are Numerical Methods and taken a Set of Polynomials we have a Set of Equations, which will provide the numeric descriptors of any phenomenon of nature, it is general. The set of polynomials by necessity always must be able to be converted into a standard analytical law in scientific notation.

We adopt the criteria that this General Mathematics of Nature Phenomena can be considered Artificial Intelligence because as a tool for Analyses, its ability for Inference, and Capacity to handle Complex Systems found in the natural world.

Artificial Intelligence is concerned with concepts and methods of inference and knowledge representation for use in making inferences. Artificial Intelligence is thought as the proper tool for the analysis of very complex systems that are difficult to model or represent. A model is a mathematics or logic representation of a prototype. A prototype is a phenomenon or assessment of a phenomenon in the real world, could be a real or theoretical prototype. Here a unique theorem is used, and the solution of any problem that can be expressed by words or numbers is found to be a selectable subset of an output set.

Selection is done by analyzing statements or polynomials within bounded limits, and selecting the subset of statements or polynomials as the model that represent the prototype. Subsets are generated by full ordered permutation of data elements; these data are words, statements or numbers. The whole of mathematics starts with a problem, and inference and deduction are used to find the solution. Here, inference and deduction is used to find the correspondence of a subset to a known problem, or the problem to a subset. The problem itself can belong to different areas of human knowledge. Modeling phenomena of high complexity or difficult analysis are readily workable.

In science and engineering, a set of polynomials generated within given bounds can provide solution to problems which otherwise need to be defined by analytic expressions. Raw data from the field or lab can form a set and subsequently an analytic expression. Uniformity of scientific and engineering reports with a simple and compact common format to substitute equations, formulas, nomographs, tables, curves or specific computer programs, is possible and shown with master samples.

PART I

Text as Data

We will start with the easier modeling, it is with words or statements as data to see how we apply our reasoning capacity to model complex systems. If we use a set of words, with the estimate of words repeated in a text, the set can reach a length of thousands of subsets, e.g., basic English have 800 words, with 10 the number of word repetition per page, then the data is 8000 elements. It is unrealistic to try to examine, at the present time, the subsets out of the set of the quantity given by the factorial of 8000. Subsets where elements are a group of words offer the easy approach to obtain logic results, but different meanings are apparent immediately, e.g.,

- The program expects action without force
- Force without action the program expects
- Expect the program without force action
- Expect force without the program action

- The integer program expects a single argument
- The integer expects a single argument program
- The integer argument expects a single program
- The argument expects a single integer program
- Program the argument single expects a integer

We were taught that syllogisms are absolutely logic. Numbers 4 and 6 in the following one fail logic:

1   all men are mortal--Socrates is a man--Socrates is mortal
2   all men are mortal--Socrates is mortal--Socrates is a man
3   Socrates is a man--all men are mortal--Socrates is mortal
4   Socrates is a man--Socrates is mortal--all men are mortal
5   Socrates is mortal--all men are mortal--Socrates is a man
6   Socrates is mortal--Socrates is a man--all men are mortal

but next modern syllogism seem to be perfect:

1   everything he likes is esoteric--no esoteric things are on TV--nothing on TV is what he likes
2   everything he likes is esoteric--nothing on TV is what he likes--no esoteric things are on TV
3   no esoteric things are on TV--everything he likes is esoteric--nothing on TV is what he likes
4   no esoteric things are on TV--nothing on TV is what he likes--everything he likes is esoteric
5   nothing on TV is what he likes--everything he likes is esoteric--no esoteric things are on TV
6   nothing on TV is what he likes--no esoteric things are on TV--everything he likes is esoteric

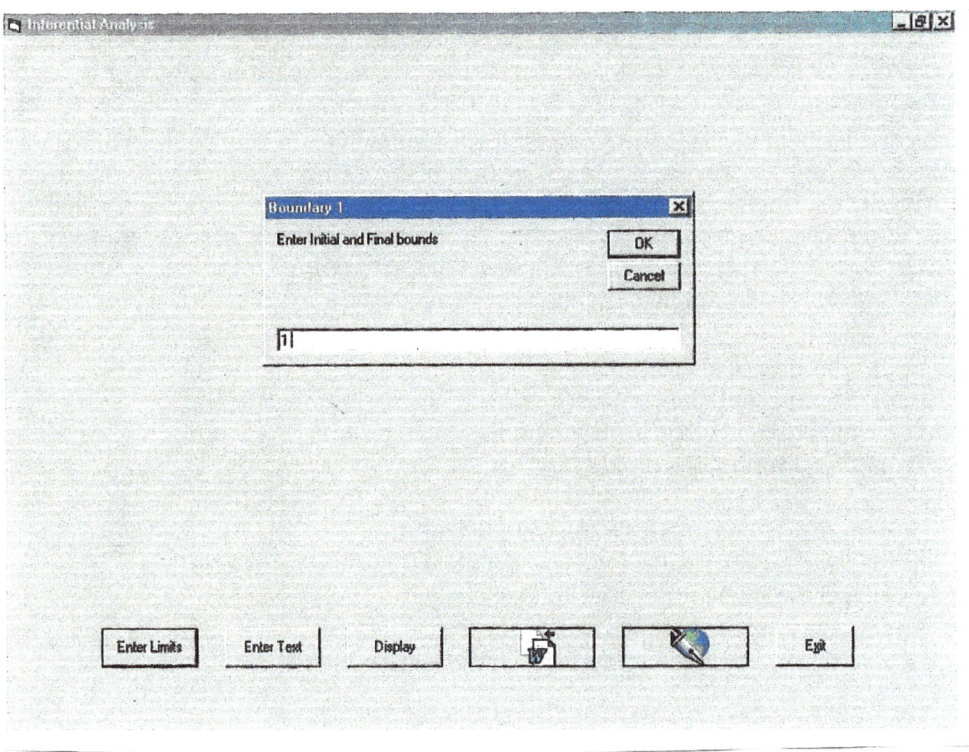

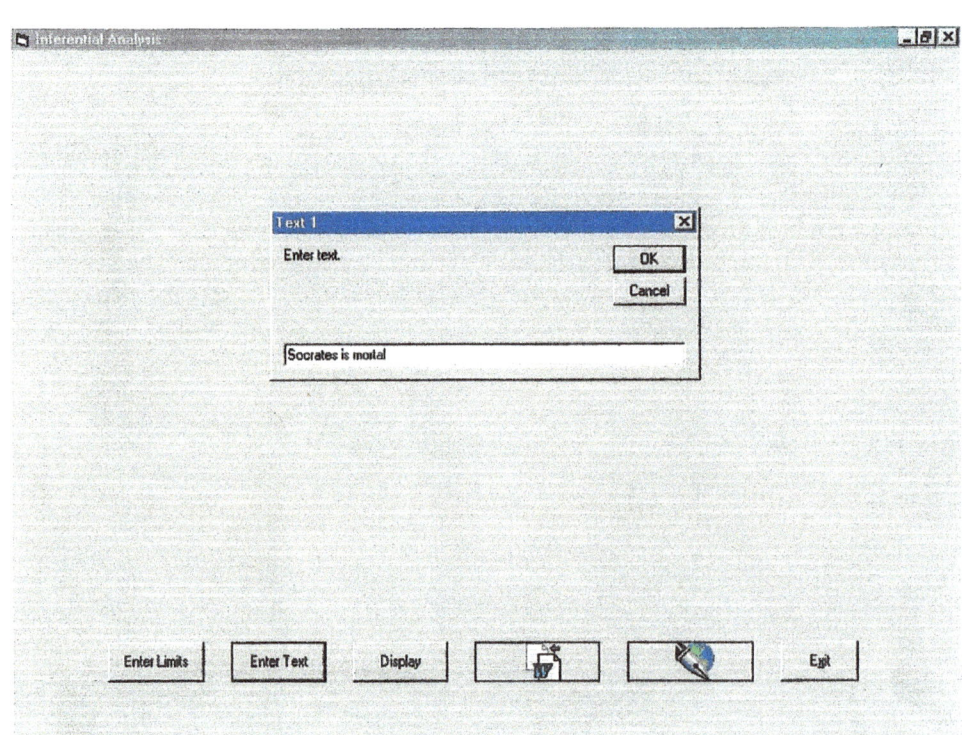

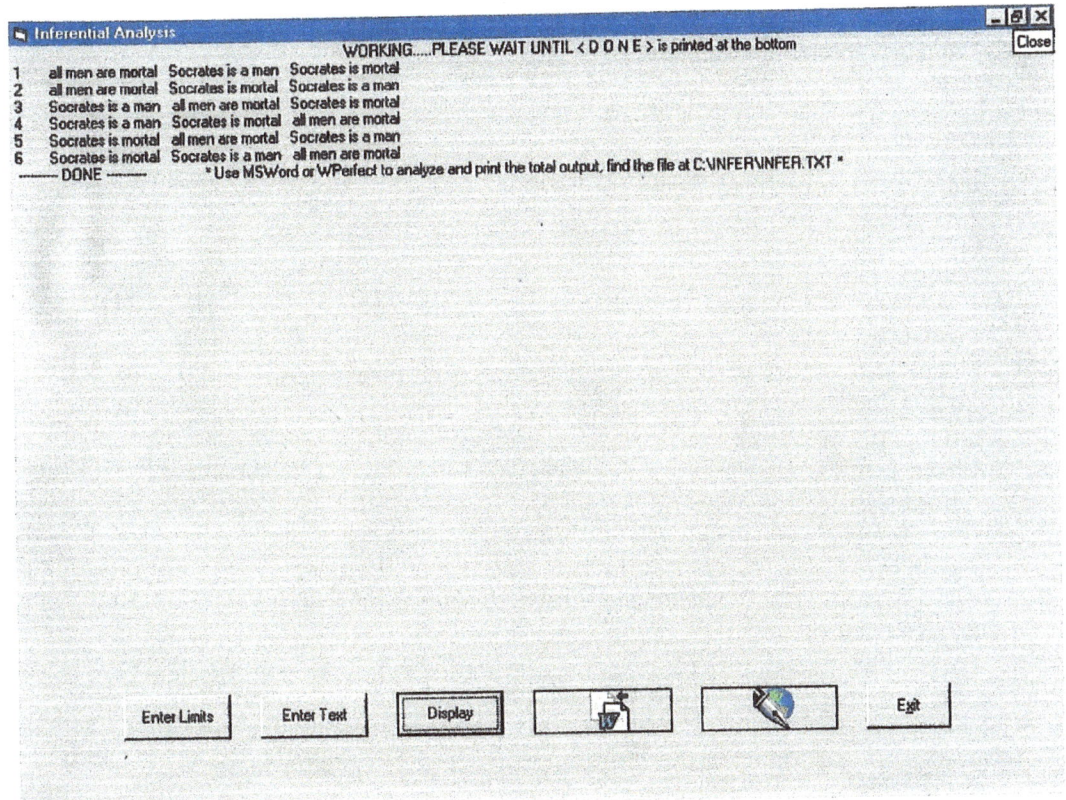

Fig. (i)

As we see in these examples we use our own mind to find logic in the set of all possibilities. In philosophy there are several methods for analysis of statements where a logic or acceptable solution is found. Here we work with the whole set of solutions, our problem is to detect a subset as a solution of a known problem, and vice versa: what are the problems that the subsets represent? The latter is going into the unknown and very exciting to make a trip to, the former requires a method to facilitate the process. There are extensive works in Artificial Intelligence with limited success, mostly because are using as data only words, and trying to imitate the human mind. We abandon such intent because its complexity and move on using the human mind as a tool analyzing pre-fabricated full ranged possibilities, which by necessity contain the solution.

The best suggestion is to use a group of words (statement) as a data element, per branch of science, literature, philosophy, or religion, as it was shown in the section of logic above, e.g., let us use conceptual data for the treatment of animal bites, where subset 14 seems to be the correct answer.

1   Immediately - Under Medical Supervision - Cleanse and Flush Wound - Antibacterial Treatment
2   Immediately - Under Medical Supervision - Antibacterial Treatment - Cleanse and Flush Wound
3   Immediately - Cleanse and Flush Wound - Under Medical Supervision - Antibacterial Treatment
4   Immediately - Cleanse and Flush Wound - Antibacterial Treatment - Under Medical Supervision
... .
12   Under Medical Supervision - Antibacterial Treatment - Cleanse and Flush Wound - Immediately
13   Cleanse and Flush Wound - Immediately - Under Medical Supervision - Antibacterial Treatment
14   Cleanse and Flush Wound - Immediately - Antibacterial Treatment - Under Medical Supervision

Probably the best advice is to define the situation in few statements, list comments / set options / write consequences, run Infer, and select from the subsets, as a list of alternatives, the logic solution for the problem.

Classical and modern prepositional logic are straightforward applications. Philosophy as thoughts expressed by words can use the text or literary approach. Even religion, if it is possible to express it by words, may have the literary approach. Logic of human intuition might be not sufficient when analyzing meaning of subsets. Examples: (1) From: I Keep Six Honest Serving-Men by Rudyard Kipling. The mood is the following:

I keep six honest serving-men
(They taught me all I knew);

Their names are What and Why and When
And How and Where and Who.
I send them over land and sea,
I send them east and west;
But after they have worked for me,
I give them all a rest.
I let them rest from nine till five,
For I am busy then,
As well as breakfast, lunch, and tea,
For they are hungry men.....

Data shows that the speaker is a good person, many subsets are of little help, but the surprise come with subset 1862:

1862   I keep six honest serving men - They taught me all I knew - Their names are What and Why and When - And How and Where and Who - I send them over land and sea - I give them all a rest - For I am busy then - I let them rest from nine till five - As well as breakfast lunch and tea - I send them east and west - For they are hungry men - But after they have worked for me -

that shows a different person able to let people die if they do not work for him. The whole sense had change, the logic is different, and a new creation appears as a character in the plot. If the author likes the second character, he can change the whole plot, or assign it to another person. From basic data and a simple run, many suggestions can be provided to the author. How many subsets will give substantial change of sense to the paragraph out of the $12! = 470,000,000$ possible subsets in a full output file? May be few subsets, it is possible to jump in time and try runs that will last as much time as you are ready to wait, this is explained in Part II for numbers but it is good for words/statements as well.

(2) From: The Gipsy Vans by Rudyard Kipling. The first part said:

Unless you come of the gipsy blood
That takes and never spares
Bide content with your given good
And follow your own affairs
(et cetera)

We make a jump in time and took as data the subset 1, and looking at subsets 19, 28, 1523 and 5039 we observe that subsets 1523 and 5039 present the character as a robber that steals by night and day!

1   Unless you come of the gipsy stock - Unless you come of the gipsy blood - Unless you carry the gipsy eyes - That see but seldom weep - That takes and never spares - That steals by night and day - Lock your heart with a double lock - Bide content with your

given good - Keep your head from the naked skies - Or the stars will trouble your sleep - And follow your own affairs - And throw the key away -

19   Unless you come of the gipsy stock - Unless you come of the gipsy blood - Unless you carry the gipsy eyes - That see but seldom weep - That takes and never spares - That steals by night and day - Lock your heart with a double lock - Bide content with your given good - And throw the key away - Keep your head from the naked skies - Or the stars will trouble your sleep - And follow your own affairs -

28   Unless you come of the gipsy stock - Unless you come of the gipsy blood - Unless you carry the gipsy eyes - That see but seldom weep - That takes and never spares - That steals by night and day -Lock your heart with a double lock - Keep your head from the naked skies - Bide content with your given good - And follow your own affairs - And throw the key away - Or the stars will trouble your sleep -

1523   Unless you come of the gipsy stock - Unless you come of the gipsy blood - Unless you carry the gipsy eyes - That see but seldom weep - That takes and never spares - Bide content with your given good - That steals by night and day - And follow your own affairs - Keep your head from the naked skies - And throw the key away - Lock your heart with a double lock - Or the stars will trouble your sleep -

5039   Unless you come of the gipsy stock - Unless you come of the gipsy blood - Unless you carry the gipsy eyes - That see but seldom weep - That takes and never spares - And throw the key away - And follow your own affairs - Or the stars will trouble your sleep - Keep your head from the naked skies - Bide content with your given good - That steals by night and day - Lock your heart with a double lock -

If we have a subset of solutions, we need to select the one that match that problem. How the work is done is presented as an example of optimization of Decision Making where inference play a mayor part, as follows:

The CEO address a Corporate meeting to analyze and take a decision regarding the following situation:

(a) A strike will occur or salaries are to be rising.
(b) A strike can produce 3M in losses in one year.
(c) Increase of salaries is 3M per year.
(d) No price increase in sales is possible.
(e) Funds are not available.

The Committee agrees that it is the situation, but no solution is found. The CEO left the room for a moment and run Infer with the underlined eight data above, due to the urgency he runs only 150 subsets, and call his attention the 36, 54, 86, 107 and 140 runs:

36   a strike will occur----salaries are to be risen----a strike can produce----increase of salaries----3M per year----funds not available----no price increase in sales----3M losses in <u>one year</u>

[Offer one year salary increase to workers, second year subject to increases in sales and/or other negotiations]

54   a strike will occur----salaries are to be rising----<u>a strike can produce</u>----<u>3M per year</u>----3M losses in one year----funds not available----no price increases in sales----increase of salaries

[By hiring non-unionized workers]

86   a strike will occur----<u>salaries are to be rising</u>----a strike can produce----no price increase in sales----3M per year----3M losses in one year----funds not available----increase of salaries

[Increase sales quantity]

107   a strike will occur----salaries are to be rising----<u>a strike can produce----funds not available</u>----increase of salaries----no price increases in sales----3M losses in one year----3M per year

[Without a strike funds are available]

140   a strike will occur----salaries are to be rising----3M losses in one year----a strike can produce----funds not available----increase of salaries----no price increases in sales----<u>3M per year</u>

[Only the first year, then other solutions]

The CEO returns to the meeting room and found discussions going on, he proposes the following options for negotiations with the Union:

1. *Offer one year salary increase to workers, second year subject to increases in sales and/or other negotiations*
2. *Increase sales quantity*
3. *Only the first year, then other solutions*

Hiring non-unionized workers, is not an option. Because without strike funds will be available, it is an option for the coming months.

The Committee agrees on this approach. It took 20 minutes. The CEO worked on suggestions, no elaborated solutions were provided, the human mind was guided to an acceptable solution.

PART II

Numbers as Data

In Part I we saw the almost infinite number of subsets that exist taking data from the real world, that is quite common, for example the vocabulary of a small book. Therefore, we must ask ourselves if that infinite exist, if it does not exist then we can look for solutions with confidence without the need to analyze an infinite set of possibilities. What follows is an analysis of this question.

If
$$\text{Lim } f(x) = 1$$
$$x \to 1$$

for x from zero to one, in 1000 steps, assuming invariability of those factors that affect f(x) in the 1000 steps. The latter is not true in natural phenomena because the variability of such factors is not instantaneous and we must introduce the factor time (t). If each step takes one year, the convergence to one, ends in 1000 years, during this time the phenomenon must be invariable. In the matter of infinite we speak of millions of years and the phenomenon cannot be invariable, at least with natural phenomena, because the environment, the background, laws of Physics like the Second Principle of Thermodynamics, and the local conditions change. For instance, if in the case of the 1000 years there is a annual change of 1/1000, at the end of 1000 years the

$$\text{Lim } f(x) = 1$$
$$x \to 1000$$

become something different, like

$$\text{Lim } f'(x) = -1$$
$$x = 1000$$

and the solution is not valid in the interval 0 < x < 1000.

In general Lim f (x) = 1 is not valid in the interval 0 < x < ∞ and

$$\text{Lim } f(x) = \text{Lim } f'(x)$$
$$x \to \infty \qquad 0 < x < \infty$$

but
$$\text{Lim } f'(x) =/= \text{Lim } f(x) \quad \text{or} \quad f(x) =/= f'(x)$$
$$0 < x < \infty \qquad x = \infty$$

then the infinite does not exist in natural phenomena because in the infinite the phenomenon is not the same. The infinite exist if, and only if, in an evolutionary process, the time needed to pass from one state to the next is zero. Mathematical functions like sin (x) are not evolutionary, except if they represent states of an evolutionary process, then the mathematical infinite exist.

Comparison with the Principle of Uncertainty (PU): The lack of precision of the PU can be solved, theoretically if the state of the technology permits, with more precise measurements, or with an appropriated change in spatial/temporal scales. Consequently, PU refers to physical phenomena that evolve with non-zero time steps, in the atomic level it is in the order of 10^{-22}.

Proof that the infinite does not exist: The relativity of space and time prove that physical phenomena is discontinuous, then the infinite does not exist. Nothing is permanent through time, then time is relative. If time is relative, then by necessity phenomena through time change and $f(x) =/= f'(x)$, then the infinite does not exist.

Is the time infinite? If time vary, then $f(t) =/= f'(t)$ and infinite time does not exist. If it is so, can we have a time void through time? The answer is not if time is continuous, but time is discontinuous because the phenomena that measure time are, then time is not infinite and a void can occur.

Is the space infinite? If space vary, then $f(s) =/= f'(s)$ and infinite space does not exist. The space is defined by fields of forces, like gravity, then the space vary and is finite.

Meaning of infinite in the physical world:

1. Mathematically the infinite is a physical impossibility
2. Physically the infinite is a mathematical impossibility
3. Physically, the infinite cannot be reached in an evolutionary process, except when the evolution ends.

Meaning of infinite in the conceptual world:

Infinite processes happen only in processes that does not evolve, example: a frozen universe without change of energy. Because the universe evolve, it is finite. Let us check with logic: If God is infinite, He is out of this universe, in a state of non-evolution (He is perfect after all). Where we can locate Him? May be superimposed to our universe? No, because our universe is finite and there is no room for an infinite, consequently He is not in our universe because He needs an infinite space. He must be found in another place where nothing evolve. What interaction can exist between an absolute entity that does not evolve, and other that evolve? The one that not evolve does not act, only the one that evolve have action but is finite in time and space, then is less strong to influence the infinite. We conclude that the logical infinite is dubious.

The following *Theorem* is enough for our purposes:

Hypothesis: For a set of numbers or words, 1 through n, any subset of the total set can be arranged exhaustively with probability one (1) of occurrence. Total possible arrangements are equal to n!

Thesis: From a set of elements, numbers or words, a subset can be extracted by means of practic or logic rules, respectively.

Proof: If the opposite is true, then the Thesis is false.

If, in the n! arrangements of the set 1 trough n elements, one is missing, then the total arrangements are (n-1)!

This is false, then the opposite is true and the certainty of occurrence is proved. All subsets that form the set exist and any of them can be found.

Axiom 1:   Most of mathematics are formed by practical rules to obtain subsets represented by numbers which provide specific results.

Axiom 2:   Most texts have logic rules to obtain comprehensible results.

Axiom 3:   Most laws of natural sciences, and works or creative arts, can be made comprehensible by subsets of words.

Axiom 4:   Most works of pictorial art are subsets of points in a line, plane or space.

Axiom 5:   An ordered set of subsets is complete when the last subset is the inverted of the first.

Examples:

All subsets of numbers (primes, logs, etc) are subsets of natural numbers.
Irrational numbers (e, pi, etc) are subsets of natural numbers.
All subsets of words are subsets of words in a language.
All pictures are subsets of pixels in a plane.
All bodies are subsets of points in space.

Set, subset, and segments of a subset are depicted in the following sketches.

```
|_____Segment →_||_____| Subset

     |_____  _____|
Set
                    .
                .
     |_____  _____|
```

a) Finite quantity of numbers or words:
```
        |_____|
```

b) Now or here start an infinite quantity of numbers or words:

```
        |_____
```

c) A quantity of numbers or words started in the infinite past or place, and finish here or now:

```
        _____|
```

d) We are here now, and a quantity of numbers or words started in the past or place, and ends in the future or some place:

_____|_____

Linear representation of a set:

|___Subset →|_|_____|_|____|_|_____|
1   n!
Ordered Set →   Location of subsets   ← Inverted Set

Example: The set 1,2,3 has n! subsets or 3! = 6, where the ordered set is 1,2,3 and the inverted set is 3,2,1. The theorem is deterministic and does not accept methods of randomness, probabilities and the like. All methods, functions, algorithms, et cetera, of mathematics are practical rules.

Segments and Set Reduction for Search Purposes:

The data file prepared has the following format (with Ranges 1 through 3, step 1):

Order	Group	Data	Range	Data x Range	Total in Row
1	1	1 2	1 1	1 2	3
2	1	1 2	2 1	2 2	4
3	1	1 2	3 1	3 2	5
4	1	1 2	1 2	1 4	5
5	1	1 2	2 2	2 4	6
6	1	1 2	3 2	3 4	7
7	1	1 2	1 3	1 6	7
8	1	1 2	2 3	2 6	8
9	1	1 2	3 3	3 6	9
10	2	2 1	1 1	2 1	3
11	2	2 1	2 1	4 1	5
12	2	2 1	3 1	6 1	7
13	2	2 1	1 2	2 2	4
14	2	2 1	2 2	4 2	6
15	2	2 1	3 2	6 2	8
16	2	2 1	1 3	2 3	5
17	2	2 1	2 3	4 3	7
18	2	2 1	3 3	6 3	9

This is an exhaustive output. Therefore, the factors of the range and the products are commutative. In this case there are two groups, one with order number 1, another with number 2.
If we want to affect each element by a constant, it is applied to the data. The same apply for values of a range. Multiplication by a constant is a particular case of multiplication by a range. The software includes the constant option as a simplification when its use is more convenient. We can reduce the

runtime by locating any subset as a segment of a set in two ways: a) By location of a specific subset in the set as function of the run number, e.g., taking 5 elements for a total of 120 subsets:

```
Run    1 =  1  2  3  4  5
Run   30 =  2  1  5  4  3   →   1/4 of total time
Run   60 =  3  2  5  4  1   →   1/2 of total time
Run   90 =  4  3  5  2  1   →   3/4 of total time
Run  120 =  5  4  3  2  1
```

b) By dividing the total subsets in $n! / (n-1)!$ parts, we find the run number and the elements in the subsets. Next table shows the process:

Shift to first Col	$7!/6! = 5040/720 = 7$ part A							N	R number (n x 5040)
1	1	2	3	4	5	6	7	$1/7 = 0.142857142857$	1
2	2	1	3	4	5	6	7	$2/7 = 0.285714285714$	721
3	3	1	2	4	5	6	7	$3/7 = 0.428571428571$	1441
4	4	1	2	3	5	6	7	$4/7 = 0.571428571429$	2161
5	5	1	2	3	4	6	7	$5/7 = 0.714285714286$	2881
6	6	1	2	3	4	5	7	$6/7 = 0.857142857143$	3601
7	7	1	2	3	4	5	6	$7/7 = 1.000000000000$	4321
								n!	5040

The software can be stopped at any run number. From this approach two rules are derived:

a) Assuming that it is impossible to wait several million years to find a particular subset, we can go directly to a segment by entering the subset given by rows A as data, defeating the time constraint and starting at the corresponding advanced run number, e.g., $50! / 49! = 3\times10E64 / 6\times10E62 = 50$ (will divide in 50 parts).

b) By using several computers, which one working a given segment, it is possible to reduce the runtime even more. A combination of a) plus b) can make a large system searchable in a period of time as short as we want, being function of the means applied.

c) Practical or logic rules. Always arrange variables for steps between extreme values (bounded limits), then find a logic rule for a quick subset extraction, e.g., max min in linear programming, principal diagonals and secondary orders in determinants and matrices, search order of words or sentences in text, close notes in music, and so on.

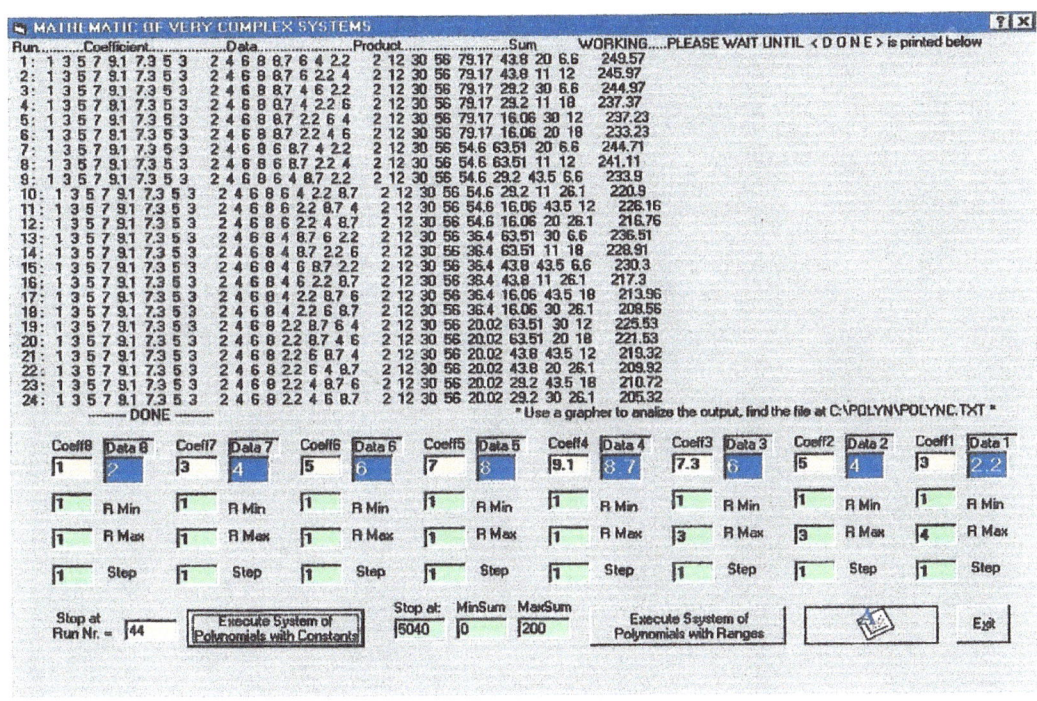

Fig. (ii), (iii)

Mathematics Setting

What is a system of polynomials? An example is the following:

$$f(x_1) = a_1 x_1^n + b_1 x_2^{n+1} + \ldots\ldots + c = 0$$
$$\ldots\ldots\ldots\ldots\ldots\ldots\ldots\ldots\ldots\ldots\ldots\ldots\ldots\ldots\ldots$$
$$f(x_m) = a_m x_1^n + b_m x_2^{n+1} + \ldots\ldots\ldots + c = 0$$

What the system means?: each polynomial represent a function, a system represent several functions horizontally. Vertically the columns of the system are relations between the variables of the polynomials. In other words, vertically is a new function between individual functions, they describe a complex system:

$f(x) = f^n(x)$, where n is the quantity of polynomials taken in the system..

Complex phenomena in physical and natural sciences can be represented by a set of polynomials, let us write a set as a system of subsets. With data: a, b, c, and d, we have:

$$ax_1 + bx_2 + cx_3 + dx_4 = A_1$$
$$ay_1 + by_2 + cy_3 + dy_4 = A_2$$
$$\ldots\ldots\ldots\ldots\ldots\ldots\ldots\ldots\ldots$$
$$az_1 + bz_2 + cz_3 + dz_4 = A_n$$

with,

$$x_i = f(x_i) = A, B, C, D$$
$$y_i = f(y_i) = A, B, C, D$$
$$\ldots\ldots\ldots\ldots\ldots\ldots\ldots$$
$$z_i = f(z_i) = A, B, C, D$$

then,

$$aA + bB + cC + dD = A_1$$
$$aA + bB + cD + cC = A_2$$
$$\ldots\ldots\ldots\ldots\ldots\ldots\ldots$$
$$aD + bC + cB + dA = A_{24}$$

The boundaries of this system are determined by the maximum and minimum of A, B, C and D as A_{max} and A_{min}. A look at solutions contained inside the boundaries is performed with as many, and as small as we want, of A, B, C, and D values. The following question is in order: If all solutions are found, which is the way to find the one that answer a specific problem? To answer with another question, this could be: Why is important to find the answer to a specific problem, which in turn, as it is defined, can be solved with a practical solution?

Practical solutions are not abundant, for instance we present solution of different problems that are solved by practical methods, like Gauss Elimination, Linear Programming, and so on; with the only purpose to prove that really we are visiting the world of solutions, but the world of solutions is itself a new world, where problems are as rare as rare are the solutions to our problems in our limited knowledge of the whole world of problems. In other words, subsets are solution to known and unknown problems. The unknown problems to which we are able to know the answers, are the subjects of our interest. Maybe that problems belong to multidimensional systems far away of our comprehension, may be that problems can open the door to solution of problems of our comprehension, may be that problems can be rewording to find a solution to an unsolved problem.

This world of solutions is not, as it must be expected by a reasonable mind, efficient in this regard, not simple, the only reward is that it consists of answers. The whole set of polynomials consists only of answers. Complex problems with no known solution are the subjects of our interest and the main aim is how to identify solutions for them. A general way is to work graphically, in other words: Set the boundaries, and with data and ranges, look at the output, then change ranges as needed until is obtained a perfect match to observed data, such that a scientific law will be the conclusion. Transform of the applicable subset(s) to a standard equation is an easy task. Most complex phenomena are represented by sequentially superimposed-picture graphs, only the last overlapping graph is seen at the end of the process. During the graph's development process the monitor looks like a movie screen while representing the data by dynamic acquisition.

We are working in the optimization of a complex system by means of the permutation of variables or, in other words, multiplying the variables by constants originally assigned to other variables of the system. In a biologic or social model means to change assignments (tasks, resources, etc) to members of the system differently to the assignment they had in the beginning.

We can ask regarding the variation in the assignment of resources: Where the maximum, minimum, and optimum are located? This is found in the data output, columns or rows, or graphically. The best is to take

$$f(x) = A x^n + B x^{n-1} + \ldots + N = Sum$$

where x have a determined value, and change the coefficients A, B,.....N. In other words, the coefficients govern the funcion and x^n gives the progression for convergence, divergence or stability. The x value is found as follows: given the products and its sum for $a_1 x^3 + a_2 x^2 + a_3 x^1 = c$

$$30 + 23.26 + 8.61 = 61.87$$

by convention we made: $\quad a_3 = 1$ or $x = 8.61$

with $x^2 = 74.13$ and $x^3 = 638.27$
we have $\quad\quad\quad\quad\quad a_2 = 23.26 / 74.13 = 0.3138$
and $\quad\quad\quad\quad\quad\quad a_1 = 30 / 638.27 = 0.047$

and finally $\quad\quad\quad 0.047 (8.61)^3 + 0.3138 (8.61)^2 + 1.0(8.61)^1 = 61.87$

The set of polynomials is a set of functions represented in a plot as a dynamic process when they converge / diverge to the final value. When the final value is reached the evolutionary process is finished and we obtain a stable state. The graph of a set of polynomials is a scatter plot or one regression equation graphed as a trend line. Any expression of that form represent a phenomenon. Actually in science and technology is used and preferred the trend line, but it is not good enough in complex system phenomena, in fact there are no general description, analytic or graphic for complex phenomena.

How to select a subset out of a set that represent better the prototype:

1. Relate the solution's parameters you have in the model with the parameters of the prototype.
2. Set the boundaries for both.
3. Check model runs outputs with known outcomes of the prototype.
4. The subset of polynomials is the final model.
5. Work with the model to make simulations of the prototype in two lines: (i) Different runs with same parameters to test behavior, and (ii) Change parameters to modify artificially the prototype (e.g.: simulations).

How to obtain an equation out of a subset:

From the spreadsheet with the data and sum at the last column, consider the columns as the x axis and the rows as the y axis, select a subset to better model the prototype (e.g.: the maximum value of the sum column) and write the regressed equation of that raw. By doing the same with any column we have the regression equation for the assignment of resources in the converging / diverging process. By transposing the whole set the situation reverses, but it is important to plot both cases to visualize the regression curves.

We can summarize as follows:

I. How to find the equivalent equation of a subset of polynomials?: With interpolated data or a regression equation.

II. How to define the representation of a complex system?: By a regression curve with minimum errors.

III. Given an equation, what is the equivalent set of polynomials?: The subset of polynomials that satisfy the equation.

When we try to find individual solutions from the data file, we must remember what they represent, namely, columns are:

$$Ax + By + C = 0$$
$$\downarrow \quad \downarrow \quad \downarrow$$

→ graph type: Figure 1

and rows are:

$$Ax + By + C = 0$$

→ graph type: Figure 2

For instance Data: 1,1 Ranges: 1 to 3 step 1 (See Table below). Conversion of a column in a row, and vice versa, is a tool when trying to find a practical rule.

	1	2	3	4	5	6	7	8	9
	Order	Group	Data 1	Data 2	Range 1	Range 2	Product 1	Product 2	Sum
1	1	1	1	1	1	1	1	1	2
2	2	1	1	1	2	1	2	1	3
3	3	1	1	1	3	1	3	1	4
4	4	1	1	1	1	2	1	2	3
5	5	1	1	1	2	2	2	2	4
6	6	1	1	1	3	2	3	2	5
7	7	1	1	1	1	3	1	3	4
8	8	1	1	1	2	3	2	3	5
9	9	1	1	1	3	3	3	3	6
10	10	2	1	1	1	1	1	1	2
11	11	2	1	1	2	1	2	1	3
12	12	2	1	1	3	1	3	1	4
13	13	2	1	1	1	2	1	2	3
14	14	2	1	1	2	2	2	2	4
15	15	2	1	1	3	2	3	2	5
16	16	2	1	1	1	3	1	3	4
17	17	2	1	1	2	3	2	3	5
18	18	2	1	1	3	3	3	3	6

The quantity of rows (polynomials) contained in a group when running POLYN with ranges are:

Rows = n! (Nr.of steps)^x where x = number of columns, or

Rows = n! . a . b . c .....   where a, b, c, ...= Nr. of steps in each column

e.g.,   Data: n = 4 (they are: 1, 2, 3, 4), then n! = 24

Range1 = 1 to 2 Step 1 (number of steps = 2)
Range2 = 1 to 2 Step 1 (number of steps = 2)
the scheme is the following:

Data:	1	2	3	4
Ranges: From			1	1
To			2	2
Step			1	1

and the quantity of rows is 24 x 2 x 2 = 96.

In the table above it is 2 (3) (3) = 18.

Algebra

The replication of some well known practical rules follow. The operations were done with Polyn. These operations and geometry output are self evident.

SUM: 2 + 2 = 4. Data: 1, 1 Constants: 2, 2

	1	2	3	4	5	6
	Order	Data 1	Data 2	Constant 1	Constant 2	SUM
1	1	1	1	2	2	4
2	2	1	1	2	2	4

SUBTRACTION: 7 - 3 = 4. Data: 1,1 Constants: 7, -3

	1	2	3	4	5	6
	Order	Data 1	Data 2	Constant 1	Constant 2	SUM
1	1	1	1	7	-3	4
2	2	1	1	7	-3	4

MULTIPLICATION: Data: 28   Constant: 12.125

	1	2	3
	Data	Constant	SUM
1	28.00	12.13	339.50

DIVISION: 24 / 4 = 6. Data: 24   Constant: 1/4 = 0.25

	1	2	3
	Data	Constant	SUM
1	24.00	0.25	6.00

POWER: $2^2 = 4$. Data: 2 Constant: 2

	1	2	3	4
	Order	Data	Constant	SUM
1	1	2	2	4
2	2	2	2	4

TABLE OF POWERS: 2^n. Data: 2, 2   Range: 2 to 32 Step 2

1	2	3	4	5	6	7
Order	Group	Data 1	Data 2	Range 1	Range 2	Product 1
1	1	2	2	2	2	4
2	1	2	2	4	2	8
4	1	2	2	8	2	16
8	1	2	2	16	2	32
16	1	2	2	32	2	64

ROOTS: $\sqrt[3]{27} = 3$

With $\sqrt[3]{a}$ $\rightarrow$ $a \left(\dfrac{1}{(\sqrt[3]{a})^{n-1}} \right) = 27 \left(\dfrac{1}{(\sqrt[3]{27})^{3-1}} \right) = 27(0.11) = 3$

Because $a = (\sqrt{1/0.11})^3 = 27$, the value 0.11 gives univocally the cubic root of 27, when our intention was to look for cubic roots. The same apply to other functions, like sin, cos, and so on.

Data: 27   Constant: 0.1112

	1	2	3
	Data	Constant	Product
1	27.00	0.1112	3.00

For $\sqrt{9} = 3$. Data: 8   Constant 0.334

	1	2	3
	Data	Constant	Product
1	9.00	0.33	3.00

For $\sqrt{64} = 8$. Data 64   Constant: 0.13

	1	2	3
	Data	Constant	Product
1	64.00	0.13	8.00

ARITHMETIC SERIE: Data: 1, 1,......,1 Constant: 3 (See Figure 3)

	1	2
	Selected Columns	
1	2	1
2	5	2
3	8	3
4	11	4
5	14	5
6	17	6
7	20	7
8	23	8
9	26	9
10	29	10

GEOMETRIC SERIE: Data: 1, 1,......., 1 Ranges: 3 to 48 Step 2 (See Figure 4)

	1	2
	Selected Columns	
1	3	1
2	6	2
3	12	3
4	24	4
5	48	5
6	96	6
7	192	7
8	384	8
9	768	9
10	1536	10

EXPONENTIAL, LOGARITHMIC, TRIGONOMETRIC AND OTHER FUNCTIONS

Use expansion, examples:

1) $e^x = 1 + (x / 1!) + (x^2 / 2!) + ... + (x^n / n!) + ...$ or $x + 0.5 x^2 + ... + (1/n!) x^n = -1$

For x = 1 $e^x = 2.7$ Data: 1,1,1,1 Ranges: 0 TO 1.1 STEP 0.1

Row 1759 provides the answer $1 + 1 + 0.5 + 0.2 = 2.7$ From the output data were deleted all except Sum = 2.7, then were selected in columns 11 and 12 the 1's values.

		2	3	4	5	6	7	8	9	10	11	12	13	14	15
	Order	G	D1	D2	D3	D4	R1	R2	R3	R4	P1	P2	P3	P4	S
1	1759	1	1	1	1	1	0.2	0.5	1	1	1	1	0.5	0.2	2.7

2) For the sin function Data: 1,1,......,1

Constants: 0.02, 0.17, 0.34, 0.50, 0.64, 0.77, 0.87, 0.94, 0.98. See Figure 5.
Even if obvious, this is a method for Inference Analysis, e.g., Figure 6 are sinusoids graphed in runs
like: Data: 1,1 Ranges: 1 to 3 Step 1

	1	2
	Selected Columns	
1	0.02	1
2	0.17	2
3	0.34	3
4	0.50	4
5	0.64	5
6	0.77	6
7	0.87	7
8	0.94	8
9	0.98	9
10	0.94	10
11	0.87	11
12	0.77	12
13	0.64	13
14	0.50	14
15	0.34	15
16	0.17	16
17	0.02	17
18	0.17	18
19	0.34	19
20	0.50	20
21	0.64	21
22	0.77	22
23	0.87	23
24	0.94	24
25	0.98	25

3) Spiral $\alpha = a.\theta$ where a is a constant and α is given in radians.

	1	2
	Selected Columns	
1	0.50	1
2	1.00	2
3	2.00	3
4	2.50	4
5	3.00	5
6	4.00	6
7	4.50	7
8	5.00	8
9	6.00	9
10	6.28	10

4) Trend and Convergence: Figure 7.
   Figure 8 is its 3D surface.

UNKNOWN FUNCTIONS

   Figure 8 used Data: 1,1,1,1,1 Ranges: 1 to 3 Step 0.5, particularly graphing columns 1,10,17 producing quite different results. There are 375,000 subsets generated in few seconds by a standard desktop computer. We observe that some group of data follows known curves defined by equations that were not used here. Sometimes we get unexpected outcomes, some can be games, one example is given by Data: 0,1,2,3,4,5,6,7,8,9 Constants: 0,0,0,0,0,0,0,1,1,1. In the output keep only those rows with the column Sum = 15 to obtain: 348, 159, 267, 249, 357, 168, 456 and 258, for rows and columns 1 through 3 and diagonals 1 and 2, to get the magic square:

4	9	2
3	5	7
8	1	6

Geometry

Figure 9: draw a octogon. Data: 3 Range: 0.25 to 2.1 Step 0.1

	1	2	3	4
	Selected Rows and Columns			
1	2	3	0.25	0.75
2	3	3	0.35	1.05
3	4	3	0.45	1.35
4	5	3	0.55	1.65
5	6	3	0.65	1.95
6	7	3	0.75	2.25
7	8	3	0.85	2.55
8	9	3	0.95	2.85
9	10	3	1.05	3.15
10	11	3	1.15	1.00
11	12	3	1.25	3.75
12	13	3	1.35	4.05
13	14	3	1.45	4.35
14	15	3	1.55	4.65
15	16	3	1.65	4.95
16	17	3	1.75	5.25
17	18	3	1.85	5.55
18	19	3	1.95	5.85
19	20	3	2.05	6.15

	1	2	3	4
		X		y
1	1	3	0.25	0.75
2	4	3	0.55	1.65
3	6	3	0.75	2.25
4	9	3	1.05	3.15
5	12	3	1.35	4.05
6	14	3	1.55	4.65
7	17	3	1.85	5.55
8	19	3	2.05	6.15
9	29	3	0.25	0.75

At this time appears that everything needs to be known just to reproduce it, that means that in fact a less efficient method is not useful. This way of thinking miss the point that we are Visiting the World of Solutions without regard to efficiency or the problem they are solving. For a circle, conic, etc. the less efficient "look to match the answer" is irrelevant, and the value of knowing only answers become obvious, e.g., a spiral and cardioid have the practical equations:

$$r = a.e^{K\theta} \qquad \text{and} \qquad r = 2\,a\,(1 + \cos\theta)$$

To exemplify how plots are obtained, three groups of graphs are included in Figures 10 through 25, they are:

Group 1: Figure 10 intentionally graph points and triangles, Data:1,2,3,4,5 Ranges: 6 to 8 Step 1 Figure 11: Data: 0.01, 100 Ranges: 1 to 0 Step 0.11 and 0 to 1 Step -0.11 and graphing column 4 vs. 9, column 5 vs. 7, column 9 vs. 1, column 5 vs. 8 and column 1 vs. 9. Figure 12 are appearances of different circles. Figures 13 through 17: something similar was obtained with Data: 3.14159 Ranges: -3 to 3 Step 0.000271, the radius is a graph of column 1 vs. column 2, the diameter are columns 4 vs. 2, the long spiral are columns 1 vs. 4, the extended paraboloid are columns 4 vs. 3, and the paraboloid + two included cardioids + extensions are graphs of columns 3 vs. 4. Figures 18 and 19: Data: 1,1 Ranges: -2 to 2 Step 0.000033333, the spirals were obtained by plotting columns 1 vs. 2, 2 vs. 4, 1 vs. 4, respectively. Other options are columns 4 vs. 3, 4 vs. 1,2,3, and 3 vs. 1,2,4. Figure 20 are graphs working with similar data, is noticeable the cardioid.

Group 2: Figure 21 is an unintentionally obtained graph. This double curvature graph was obtained with Data: 1 Range: 20 to 0. Step -1 graphing column 1 vs. 3 or vice versa. Figures 22 and 23 show the hidden geometric properties out of a data file, which graphed in Cartesian 2D shows no indication of them. Data: 1, -1, 0.5, -1 Ranges: 7 to 8 Step 0.2, 8 to 7 Step -0.2, 7 to 8 Step 0.2, and 8 to 7 Step -0.2, showing a graph that exceeds the limits of the polar boundaries, using column 15 vs. 1, and using column 13 vs. 15 we see the appearance of evolvents, exceeding the limits of the polar boundaries and even the boundaries of the paper size, with properties of chaotic behavior. Figures 22 and 23 exceeds the limits of the monitor. Figure 24 is a 3D view of the same, it is an example of complex topography.

Group 3: Search for geometric patterns. Figure 25: Data: 1, 5, 10, 14 Ranges: 1 to 2 Step 0.3, graphing several combinations of columns.

Geometric Game: Figure 25 is particularly interesting, appear at the center the solution of a suggested problem: "Draw two superimposed pentagons, using nine lines, without rising the pen."

The path is the following:

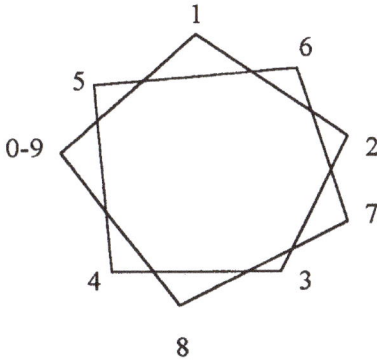

Matrices and Determinants

Matrices and determinants are based in practical rules, which can be one of the following in a 3 x 3 case:

Mathematicians adopted #5. Because a 3 x 3 have 9 elements, any option is a subset of the 9 elements set.

Examples:

1)
$$\begin{vmatrix} 1 & 2 \\ 3 & 4 \end{vmatrix} = (1 \times 4) - (2 \times 3) = 4 - 6 = -2$$

Data: 0,1,2   Constants: 4, -3

	1 Group	2 D1	3 D2	4 D3	5 P1	6 P2	7 SUM
1	0	1	2	0	4	-6	-2
2	0	2	1	0	8	-3	5
3	1	0	2	1	0	-6	-5
4	1	2	0	1	8	0	9
5	2	0	1	2	0	-3	-1
6	2	1	0	2	4	0	6

With Ranges: -2 to 1 Step 1 we obtain the same result. Note that the row with data and constants in proper order gives us the column SUM as solution.

	1 GROUP PP	2 D1	3 D2	4 D3	5 R1	6 R2	7 R3	8 P1	9 P2	10 P3	11 SUM
1	1	0	4	3	-2	1	-2	0	4	-6	-2
2	1	0	4	3	-1	1	-2	0	4	-6	-2
3	1	0	4	3	0	1	-2	0	4	-6	-2
4	1	0	4	3	1	1	-2	0	4	-6	-2
5	1	0	4	3	-2	0	-1	0	0	-3	-3
6	1	0	4	3	-1	0	-1	0	0	-3	-3

2)   1   2   3

    4   5   6 $= (1 \times 5 \times 6)+(2 \times 6 \times 7)+(3 \times 4 \times 8)-(1 \times 6 \times 8)-(2 \times 4 \times 9)-(3 \times 5 \times 7)=$
                $= 30+84+96-48-72-105 = -15$

    7   8   9

First row is data, all others are constants. Data: 0,1,2,3,1,2,3 Constants, 30,42,32,-48,-36,-35

	1 D1	2 D2	3 D3	4 D4	5 D5	6 D6	7 D7	8 P1	9 P2	10 P3	11 P4	12 P5	13 P6	14 P7	15 S
1	0	1	2	3	1	2	3	0	30	84	96	-48	-72	-105	-15
2	0	1	2	3	1	3	2	0	30	84	96	-48	-108	-70	-16
3	0	1	2	3	2	1	3	0	30	84	96	-96	-36	-105	-27
4	0	1	2	3	2	3	1	0	30	84	96	-96	-108	-35	-29
5	0	1	2	3	3	1	2	0	30	84	96	-144	-36	-70	-40

Use Range if the result is unknown.

1.

$$\begin{vmatrix} 6 & 9 & 15 \\ 3 & 7 & -2 \\ 16 & 14 & 1 \end{vmatrix}$$ = (6x7x-2)+(9X-2X16)+(15x3x14)-(16x14x-2)-(9x3x1)-(15x16x7) =
= -84-288+630+168-27-1680 = -1281

Data:0,6,9,15,6,9,15 Constants: -14,-32,42,28,-3,-112

	1	2	3	4	5	6	7	8	9	10	11	12	13	14	15
	D1	D2	D3	D4	D5	D6	D7	P1	P2	P3	P4	P5	P6	P7	SUM
1	0	6	9	15	6	9	15	0	-84	-288	630	168	-27	-1680	-1281
2	0	6	9	15	6	15	9	0	-84	-288	630	168	-45	-1008	-627
3	0	6	9	15	9	6	15	0	-84	-288	630	252	-18	-1680	-1188
4	0	6	9	15	9	15	6	0	-84	-288	630	252	-45	-672	-207
5	0	6	9	15	15	6	9	0	-84	-288	630	420	-18	-1008	-348
6	0	6	9	15	15	9	6	0	-84	-288	630	420	-27	-672	-21
7	0	6	9	6	15	9	15	0	-84	-288	252	420	-27	-1680	-1407
8	0	6	9	6	15	15	9	0	-84	-288	252	420	-45	-1008	-753
9	0	6	9	6	9	15	15	0	-84	-288	252	252	-45	-1680	-1593
10	0	6	9	6	9	15	15	0	-84	-288	252	252	-45	-1680	-1593
11	0	6	9	6	15	15	9	0	-84	-288	252	420	-45	-1008	753

Simultaneous Equations

1. Polynomials

$$P_N(x) = a_0 + a_1 x + a_2 x^2 + \ldots\ldots\ldots + a_N x^N = \sum_{K=0}^{N} a_K x^K$$

Use matrix:

$$\begin{vmatrix} y & 1 & x \ldots\ldots\ldots\ldots\ldots x_N \\ y_2 & 1 & x_2 \ldots\ldots\ldots\ldots x_2^{N} \\ \ldots & \ldots & \ldots & \ldots \\ y_{N+1} & 1 & x_N \ldots\ldots\ldots\ldots x_{N+1}^{N} \end{vmatrix}$$

2. Gauss Elimination

a) 3x - 4y + 1z = 1
b) 2x +1y + 2z = 3
c) 1x + 2y - 1z = 5

a) Data: 3,-4,1 Ranges: -1 to 4 Step 1. Find order 3, -4, 1 with Sum = 1

$3x - 4y + 1z = 1$

	1	2	3	4	5	6	7	8	9	10
	a	B	c	x	y	z				
1	3	-4	1	2	1	-1	6	-4	-1	1
1	3	-4	1	-1	-1	0	-3	4	0	1
1	3	-4	1	3	2	0	9	-8	0	1
1	3	-4	1	0	0	1	0	0	1	1
1	3	-4	1	4	3	1	12	-12	1	1
1	3	-4	1	1	1	2	3	-4	2	1
1	3	-4	1	2	2	3	6	-8	3	1
1	3	-4	1	-1	0	4	-3	0	4	1
1	3	-4	1	3	3	4	9	-12	4	1

b) Data: 2,1,2   Ranges: -1 to 3 Step 1. Find order 2,1,2 with Sum = 3

$2x + 1y + 2z = 3$

	1	2	3	4	5	6	7	8	9	10
				x	y	z				
1	2	1	2	3	-1	-1	6	-1	-2	3
1	2	1	2	2	1	-1	4	1	-2	3
1	2	1	2	1	3	-1	2	3	-2	3
1	2	1	2	2	-1	0	4	-1	0	3
1	2	1	2	1	1	0	2	1	0	3
1	2	1	2	0	3	0	0	3	0	3
1	2	1	2	1	-1	1	2	-1	2	3
1	2	1	2	0	1	1	0	1	2	3
1	2	1	2	-1	3	1	-2	3	2	3
1	2	1	2	0	-1	2	0	-1	4	3
1	2	1	2	-1	1	2	-2	1	4	3
1	2	1	2	-1	-1	3	-2	-1	6	3

c) Data: 1,2,-1   Ranges: -1 to 3 Step 1. Find order 1,2,-1 with Sum = 5

$1x + 2y - 1z = 5$

	1	2	3	4	5	6	7	8	9	10
				x	y	z				
1	1	2	-1	2	1	-1	2	2	1	5
1	1	2	-1	0	2	-1	0	4	1	5
1	1	2	-1	3	1	0	3	2	0	5
1	1	2	-1	1	2	0	1	4	0	5
1	1	2	-1	-1	3	0	-1	6	0	5
1	1	2	-1	2	2	1	2	4	-1	5
1	1	2	-1	0	3	1	0	6	-1	5
1	1	2	-1	3	2	2	3	4	-2	5
1	1	2	-1	1	3	2	1	6	-2	5
1	1	2	-1	2	3	3	2	6	-3	5

In the three runs, x, y, and z have the common values x = 2, y = 1, and z = -1, that is the solution.

	1	2	3	4
	x	y	z	SUM
1	2	2	1	5
2	6	-4	-1	1
3	4	1	-2	3

Second Degree Equation

$$ax^2 + bx = c$$
$$2x^2 + 4x = -2$$

Data: 0,2,4 Constants: 0, +1, -1 (remember that $(-1)^2 = +1$. Solution: x = -1

	1	2	3	4	5	6	7
	D1	D2	D3	P1	P2	P3	SUM
1	0	2	4	0	-2	-4	-6
2	0	4	2	0	-4	-2	-6
3	2	0	4	2	0	-4	-2
4	2	4	0	2	-4	0	-2
5	4	0	2	4	0	-2	2
6	4	2	0	4	-2	0	2

With Ranges: -1 to 1 Step 0.5. Identify row 2,4,0 that shows 2, -4,0, -2 for the solution.

	1	2	3	4	5	6	7	8
	Group	D1	D2	R1	R2	P1	P2	SUM
1	1	2	4	0.50	-1.00	1.00	-4.00	-3.00
2	1	2	4	1.00	-1.00	2.00	-4.00	-2.00
3	1	2	4	-0.50	-0.50	-1.00	-2.00	-3.00
4	1	2	4	0.00	-0.50	0.00	-2.00	-2.00
5	1	2	4	0.50	-0.50	1.00	-2.00	-1.00
6	1	2	4	1.00	-0.50	2.00	-2.00	0.00
7	1	2	4	-1.50	0.00	-3.00	0.00	-3.00

For a cubic equation: $2x^3 + 3x^2 + 80x = 0$ the data is 2,3,80 and, with constants or ranges, find $x = v$ such that $v^2 = x^2$ and $v^3 = x^3$, and similarly for higher degrees.

Differentiation and Integration

Differential Equations: $\qquad$ $y^{[\cdot]} = 1/4\,(3.1416)^2\,(x - y)$ at $x = 5$
subject to the boundary conditions: $y(0) = 1$ and $y(1) = 1$
Solution: $y = x + \cos(1/2\,(3.1416).\; x = 5 + \cos(5/2(3.1416)) = 5 + \cos(7.8539) = 5 + 0.99 = 5.99$
Data: 5,1. Ranges: 1 to 2 Step 1, 0.6 to 1.0 Step 0.01. One run is enough.
In other cases it is possible to solve with matrices, e.g., systems of second order constant-coefficient equations.

	1	2	3	4	5	6	7	8	9
	Order	Group	D1	D2	R1	R2	P1	P2	SUM
1	20	1	5	1	1	1.00	5	1.00	6.00
2	19	1	5	1	1	0.99	5	0.99	5.99
3	18	1	5	1	1	0.98	5	0.98	5.98
4	17	1	5	1	1	0.97	5	0.97	5.97
5	16	1	5	1	1	0.96	5	0.96	5.96

Integration: Data: 1,1,........,1. Constants: 1.5,4,6,6.5,5,2. Integration is given as Sum = 25, this value can be matched with the practical Simpson's approach.

Linear Programming is in general solved by the Simplex Method, we will solve equally the following sample problem:

$$4.5\,x + 4\,y \qquad = M$$
$$30\,x + 12\,y < \text{or} = 6000$$
$$10\,x + \;\;8\,y < \text{or} = 2600$$
$$\;\;4\,x + \;\;8\,y < \text{or} = 2000$$

the Simplex Method provides the solution:

```
Final Solution reached after 3 pivots.
Maximum value of objective function = 1250
variable                value
  X 1                 100.000000
  X 2                 200.000000
slack                   value
  S 1                 600.000000
  S 2                    .000000
  S 3                    .000000
constraint            shadow price
  C 1                    .000000
  C 2                    .41666667
  C 3                    .08333333
```

where x = 100, y = 200 and M = 1250 with 6000 with a slack value of 600, meaning 5400, then:

1) For 4 x + 8 y: Data: 0,4,8, Range: 90 to 220 Step 10. Select all Sum = 2000

	1	2	3	4	5	6	7	8	9	10
Group	D1	D2	D3	R1	R2=x	R3=y	P1	P2	P3	SUM
1	0	4	8	100	100	200	0	400	1600	2000
1	0	4	8	110	100	200	0	400	1600	2000
1	0	4	8	120	100	200	0	400	1600	2000
1	0	4	8	130	100	200	0	400	1600	2000
1	0	4	8	140	100	200	0	400	1600	2000
1	0	4	8	150	100	200	0	400	1600	2000
1	0	4	8	160	100	200	0	400	1600	2000
1	0	4	8	170	100	200	0	400	1600	2000
1	0	4	8	180	100	200	0	400	1600	2000
1	0	4	8	190	100	200	0	400	1600	2000
1	0	4	8	200	100	200	0	400	1600	2000
1	0	4	8	210	100	200	0	400	1600	2000
1	0	4	8	220	100	200	0	400	1600	2000

2) For 10 x + 8 y: Data: 0,10,8, Range: 90 to 220 Step 10. Select all Sum = 2600

	1	2	3	4	5	6	7	8	9	10
Group	D1	D2	D3	R1	R2=x	R3=y	P1	P2	P3	SUM
1	0	10	8	90	100	200	0	1000	1600	2600
1	0	10	8	100	100	200	0	1000	1600	2600
1	0	10	8	110	100	200	0	1000	1600	2600
1	0	10	8	120	100	200	0	1000	1600	2600
1	0	10	8	130	100	200	0	1000	1600	2600
1	0	10	8	140	100	200	0	1000	1600	2600
1	0	10	8	150	100	200	0	1000	1600	2600
1	0	10	8	160	100	200	0	1000	1600	2600
1	0	10	8	170	100	200	0	1000	1600	2600
1	0	10	8	180	100	200	0	1000	1600	2600
1	0	10	8	190	100	200	0	1000	1600	2600
1	0	10	8	200	100	200	0	1000	1600	2600
1	0	10	8	210	100	200	0	1000	1600	2600
1	0	10	8	220	100	200	0	1000	1600	2600

3) For 30 x + 12 y: Data: 0,30,12, Range: 90 to 220 Step 10
There are not x = 100 and y = 200, where the Sum is 6000 but 5400

| Group | 1 | 2 | 3 | 4 | 5 | 6 | 7 | 8 | 9 | 10 |
	D1	D2	D3	R1	R2=x	R3=y	P1	P2	P3	SUM
1	0	30	12	90	100	200	0	3000	2400	5400
1	0	30	12	100	100	200	0	3000	2400	5400
1	0	30	12	110	100	200	0	3000	2400	5400
1	0	30	12	120	100	200	0	3000	2400	5400
1	0	30	12	130	100	200	0	3000	2400	5400
1	0	30	12	140	100	200	0	3000	2400	5400
1	0	30	12	150	100	200	0	3000	2400	5400
1	0	30	12	160	100	200	0	3000	2400	5400
1	0	30	12	170	100	200	0	3000	2400	5400
1	0	30	12	180	100	200	0	3000	2400	5400
1	0	30	12	190	100	200	0	3000	2400	5400
1	0	30	12	200	100	200	0	3000	2400	5400
1	0	30	12	210	100	200	0	3000	2400	5400
1	0	30	12	220	100	200	0	3000	2400	5400

and the Maximum value of the objective function is M = 4.5 (100) + 4 (200) = 450 + 800 = 1250

Nonlinear Programming

In general, nonlinear programming is to find the extrema of a performance measure which is nonlinear real valued function of n parameters. To solve the problem, are applied direct methods like a search technique. The Fibonacci search uses the numbers of tests to be made (N), the width of the initial interval of uncertainty ($c_2 - c_1$) that is (d_1) and the width of the interval of uncertainty after k tests have been made (d_k), where $F_N = d_1 / d_N$

$$F_N = F_{N-1} + F_{N-2} \quad \text{and} \quad F_0 = F_1 = 1$$

Data: 1,1,.....,1. Range: 1 to 90 Step 1. Select those which are the sum of the previous two. They

are: 1   2   3   5   8   13   21   34   55   89

The reduction ratio d_N / d_1 is the inverse of the same numbers, or:

1   0.5   0.333 0.2   0.125 0.077 0.047 0.029 0.018 0.011

	1	2	3
1	1	1	1.00
2	2	2	0.50
3	3	3	0.33
4	4	5	0.20
5	5	8	0.13
6	6	13	0.08
7	7	21	0.05
8	8	34	0.03
9	9	55	0.02
10	10	89	0.01

Figure 26 is a plot of column 1 versus column 2.

Dynamic Programming

Dynamic Programming can be viewed as a succession of decision problems, each one building on the last, until the problem is solved. Our approach will be to provide possibilities of occurrence for the variables and solve the problem at once, valid for a given time only. Examples are given later. Dynamic Programming can be applied to a wide variety of different problems, both linear and nonlinear, but must be characterized in terms of a Markovian process. This process has the property that after any numbers of decisions, k, the effect of the remaining stages depends only upon the state of the system at the end of the kth decision and the subsequent decisions. Rather than store all computed values of a maximum return function, as is done in a straightforward dynamic programming solution, many types of series approximations may be used, the most commonly used being polynomial approximation, then polynomials can be used to solve these problems.

Finite Differences and Finite Elements: In the context of numerical methods for differential equations of first and second order, a key step is to approximate the first and second derivatives by using finite differences. The errors in such formulae are typically determined by using Taylor series, called Taylor Polynomials, then solved as a polynomial. The same apply to finite elements, although matrices are frequently applied, that in time we solve with polynomials.

Probabilities and Statistics

Finding the mean is an example: Data: 3,9,12   Constants: 1/(3/3) , 1/(9/3), 1/(12/3) equal to 1, 0.3334, 0.25. Result = 24/3 = 8

	1	2	3	4	5	6	7
	D1	D2	D3	P1	P2	P3	SUM
1	3	9	12	1	3	4	8

All statistics are a consequence in the use of a practical equation or formula, and need to be treated as such. Probabilities are part of dynamic programming in many social and natural sciences.

Fractals and Chaos

Some important characteristics of these new branches of mathematics are symmetry, scaling and attractors. We will try to find this characteristics. Back feeding is a typical practical method, e.g.,

$$x(I) = y(I-1)$$
$$y(I) = -(A-1) * x(I-1) + y(I-1) - (A*x(I-1) * y(I-1))$$

to see results, let us make:

$$x(I) = x(I) * 200$$
$$y(I) = y(I) * 200 \qquad \text{and we get:}$$

Order	x	y
1	1	1
2	1	1.8
3	1.8	2.52
4	2.52	3.6864
5	3.6864	5.025427
6	5.025427	6.490613
7	6.490613	7.751687
8	7.751687	8.561918
9	8.561918	8.901506
10	8.901506	8.985835

Data: 1 Range: 1 to 10 Step 0.01

	1	2	4	5
	Selected Rows and Columns			
1	1	1.00	1.00	1.00
2	81	1.00	1.00	1.80
3	153	1.00	1.80	2.52
4	269	1.00	2.52	3.68
5	403	1.00	3.68	5.02
6	550	1.00	5.02	6.49
7	676	1.00	6.49	7.75
8	757	1.00	7.75	8.56
9	791	1.00	8.56	8.90
10	799	1.00	8.90	8.98
11	-	1.00	8.98	10.00

Figure 27 shows some tendency to an attractor even with the few data taken. Because the back feed is observed to be the displacement of a row from two adjacent columns. For instance, the following data can be used:

	1	2	3	4	5
		Selected Rows and Columns			
1	1	1.50	1.50	2.00	2.50
2	2	1.50	2.00	2.50	3.00
3	3	2.00	2.50	3.00	3.50
4	4	2.50	3.00	3.50	4.00
5	5	3.00	3.50	4.00	4.50
6	6	3.50	4.00	4.50	5.00
7	7	4.00	4.50	5.00	5.50
8	8	4.50	5.00	5.50	6.00
9	9	5.00	5.50	6.00	6.50
10	10	5.50	6.00	6.50	7.00

For specific symmetries we can prepare many data files, e.g.,

1) Data: 1.30, 3.70, -1.1, 0.67 Ranges: 0 to 1 Step 0.5
   Graph of columns 12,13,14 are Figures 28 and 29
   Graph of columns 10,11,12 is Figure 30 in 2D
   Graph of columns 10,11,12 is Figure 31 in 3D

2) Data: 1,1 Ranges: -1 to 1 Step -0.2 and 1 to -1 Step 0.2
   Figure 32 shows a 2D symmetric graph. Figure 33 shows the same downscaled.

3) If we want to go a little further in degree of complexity, we can try the following:
   4.1) Data: 3,7 Ranges: -1 to 1 Step 0.2 Columns 4,6,8 Figures 34 shows symmetry
   4.2) Data: 0.7,-0.3,4 Ranges: 1 to 3 Step 0.1 Columns 9,10,11 shows attraction
   4.3) Data: 1,1,1 Ranges: 1 to 2 Step 0.3

Figure 35 is a good example of scaling while a process evolve Data: 3.14159, 2.7172 Ranges: -1 to 1 Step 0.01234 and 1 to -1 Step -0.04321, with boundaries in +2 and -2, taking only one run.

One important observation about the capacity of our human brain and level of development: we obtain recognizable geometry and the algorithms/calculus they represent, but with more than two elements, our capacity to recognize is limited, and we can speak about recognition of patterns but not specifics because we did not reach the mathematic level to identify a pattern with a specific problem, simply the solution stand alone and we use it for enjoyment, like in the pictorial art and to help as much as possible in the creative literature, logic analysis, managerial and strategic matters, games and optimization's problem.

Inference of Problem from Solution

In general, with a trend line a least square approximation is acceptable, with known selected points, but the same approximation can be reached with a different group of points. By taken all subsets of a set in the approximation seems that the data points are better correlated, and different from the one obtained as the least square approximation. Which one is the best? The best is the perfect match between the plotted data and observed data. The plotted approximation is about the unknown, the observed data is known, then it is necessary to plot a set in which some of the subsets are as close as possible to the observed data. Under such conditions check in each case what is best, the least squares approach or subsets which form a set or group of simple related elements? We can sketch a polynomial fit for the line defined by three points (1,1 - 2,3 - 3,2), taken these three points or subsets like: Data: 1,1 Ranges: 1 to 3 Step 1 Columns 1,5,6.

	1	2	3	4	5	6	7	8	9
	Order	Group	D1	D2	R1	R2	P1	P2	SUM
1	1	1	1	1	1	1	1	1	2
2	6	1	1	1	3	2	3	2	5
3	8	1	1	1	2	3	2	3	5

The approach of taken all subsets of the set that includes the observed data, provides an exact solution to this problem. There will be not a line, but a region. The need to choose the proper boundaries, and range steps for proper precision, requires to develop skills in this computational area.

Planning of experiments and induction (extract a law) from experimental data, may have the following format:

(1) Take the data A, B, C for a result D, and

(2) Adopt the data A, B, C and apply constants or ranges till get D as a Sum out of a family of subsets.

(3) This implies that everything is reduced to sums as shown in the fundamental operations, and a law can be expressed as DATA....+ RANGES....+ SELECT....Because this software handle 1 through 8 elements, and data can be all ones, ranges in fact mean boundaries, and everything rest in selecting between boundaries. For text the range is the number of elements (words, paragraphs, or group of words).

Examples are given at the end of Part III.

PART III

Real Life Applications

Physics, Chemistry

It is useless to repeat the graph representation of known laws of physics, when already has been shown the fundamental mathematical operations, with which they are solved. Laboratory planning and data analysis are self explanatory. Therefore, two main fields could utilize the wider use of subsets, they are: (1) Find the law of unknown data as Data+Ranges+Select, e.g., in the high energy particles field defines a law for an observed trajectory, where the laws of the causing forces are unknown. (2) Find the law of complex systems, taken subsets of the set that include observed data by the Data+Ranges+Select approach. The opposite of (1) could be: What interactions make for certain the existence of an undetected (undetectable) particle? Note no words like combination of interactions, or most probable existence, were used. The opposite of (2) could be: From measured data, to what determinable environment they may belong? For these difficult problems, most likely using the Infer and Polyn programs will provide a quicker path to the solution. Which in due time can be expressed in a traditional, classical, practical approach, so far the boundaries are defined and error/precision is acceptable.

Chemistry encompasses all of the above plus specifics of this branch of science, e.g., Figure 36 shows how complex chemical bonding could be defined as a subset by a law of the form Data + Ranges + Select. For Figure 36 use Data: 1,2,3,4,5,6,7 and graphing columns 13,14 and15. The same in biological developments. Some outputs in 3D resemble crystal growth and electron microscopic view of metallic surfaces.

Biology

An example of the basic Predator-Prey problem: Predators' 1, 2 and 3 prey 20%, 30% and 50% of the available resources, the total resources are T = (1x0.2) + (2x0.3) + (3x0.5) = 2.3
Data: 1,2,3 Constants: 0.2, 0.3, 0.5

	1	2	3	4	5	6	7	8
	Order	D1	D2	D3	P1	P2	P3	SUM
1	1	1	2	3	0.30	0.40	1.50	2.20
2	2	1	3	2	0.30	0.60	1.00	1.90
3	3	2	1	3	0.60	0.20	1.50	2.30
4	4	2	3	1	0.60	0.60	0.50	1.70
5	5	3	1	2	0.90	0.20	1.00	2.10
6	6	3	2	1	0.90	0.40	0.50	1.80

We obtain a minimum of 1.7 to a maximum of 2.3, for the survival of the three species, these values assume that any of them can survive with resources within the limits of 0.2 through 0.5 of the total resources available. Most of the biological models can be represented by graph showing the solution of a system of differential equations, which spirals in toward a critical point, very much like an attractor. It seems that genetic information pass through a step-wise process of development, very much like a range evolution in the generation of a column out of many subsets, namely coefficients of polynomial, where the sets or full polynomials (rows) are intermediate step-wise governing laws. This can be visualized in Figures 37 through 40, where complex views are developing, each one covering the previous one, likewise the formation of intermediate parts up to a full living body, when only the external coverage is seen at the end, this process is seen in the monitor, the figures are the last stages. Column data may well be the way DNA division proceeds, one step at a time, one arrangement at a time controlled by a step related to a quantum of energy, or a multiple of it. A practical or logical rule in this process may be well-shed a light on the complexity of life itself. The Figures show a captured sequence of events with data: 10, -10, 20, -20 and ranges: 2, 1, -0.2; -1, 2, 0.2; 2, 1, -0.2, and -1, 2, 0.2.

The resemblance to the development of a live body is as follow: First of all, from the center emanate everything recalling orders from an infinitesimal point like genes or central nervous system. Figure 37 resemble the development of a supporting skeleton or limits of assemblage, Figure 38 the development of internal organs and connectivities, Figure 39 the completion of a functioning body and the growth of a protective cover and Figure 40 looks like a final external shell. It is important to say that all this is a resemblance, but looks like a good math model for further input of data from a genome (not necessarily the most complex human one), to understand how life develops and works.

Environment

Practically all other sciences apply to environmental problems, like physics, chemistry, biology, agronomy and so on. For changes in the environment, naturally occurring or provoked requires the analysis of options. It is a system very complex, because it is necessary to define an equilibrium state and what changes will let the system to pass to another complex system in equilibrium, desirable or undesirable. Even under scarcity of data, it is possible to select these data from subsets, such that a passage from one state of equilibrium to another can be defined by a law of the type Data+Ranges+Select.

Medicine, Psychology

For experimental medicine, what was mention for physics applies. Equally apply analysis and adjustment for lab data or epidemiological data. The most promising area is the analysis of alternatives in experiment planning, combination of treatments for max-min results, and theoretical analysis of microbiology interactions, in particular analysis of viruses behavior. Combined use of Polyn and Infer programs must be used. In Psychology the use of Infer programs is likely to be more widely applicable for analysis and Polyn for tests.

Example: Pills must be given to patients each 6 hours, and are available in 6 mg, 4 mg, and 8 mg. One, two, or three pills can be taken each time. What quantity of pills are needed each time such that the average intake will be 8 mg. , and a minimum of 4 mg. must always be in the patient's system?

Data: 1,2,3 Constants: 6, 4, 8. The output is the following:

	1	2	3	4	5	6	7	8
	Interval	D1	D2	D3	P1	P2	P3	SUM
1	1	1	2	3	6	8	24	38
2	2	1	3	2	6	12	16	34
3	3	2	1	3	12	4	24	40
4	4	2	3	1	12	12	8	32
5	5	3	1	2	18	4	16	38
6	6	3	2	1	18	8	8	34

Figure 49 is the graph of the 6-hour interval vs. P1 showing an increasing concentration of the active substance in the patient's system. Figure 50 is the graph of the 6-hour interval vs. P2 showing a regular variation around the average concentration of the active substance in the patient's system. Figure 51 is the graph of the 6-hour interval vs. P3 showing a decreasing concentration of the active substance in the patient's system. P2 is the best option. The average is 2 pills times 4 mg, or 8 mg. At the 6 hrs interval will be taken pills of 4 mg. in the following quantities: 2 ,3, 1, 3, 1, 2....2, 3, 1, 3, 1, 2, and so on. In 36 hours a total of 48 mg. are taken, that is 12 pills times 4 mg. Usually three pills could be given, to be sure that will always be active substance in the patient's system, in such a case in 36 hours a total of 72 mg. are taken, that is 3 pills times 6 hrs times 4 mg. The excess of drugs ingested is 72 - 48 = 24 mg. or 50% more than needed. For those drugs with secondary effects this optimization could be of prime importance, the same apply for costly medicaments.

With x = D1, y = D2, z = D3, the set of polynomials used was:

$$
\begin{aligned}
& \quad\quad\; \mathbf{B} \\
& \quad\quad\; \downarrow \\
& 6x + 8y + 24z = 38 \\[4pt]
& 6x + 12y + 16z = 34 \\[4pt]
& \underline{12x + 4y + 24z = 40} \;\rightarrow\; \mathbf{A} \\[4pt]
& 12x + 12y + 8z = 32 \\[4pt]
& 18x + 4y + 16z = 38 \\[4pt]
& 18x + 8y + 8z = 34
\end{aligned}
$$

Row A is the equation for maximum active substance. Column B is the time-dependent coefficients of the system that optimize the dose according the constraints/boundaries/precision requested.

Technology, Architecture

Technology uses most of the sciences that deal with hardware design and construction, and easily can apply the cases described in several ways. The Critical Path Method used in management, being a max-min problem can be easily duplicated, even if not efficiently. Nonetheless, what the CPM cannot normally do, like continuous maximums monitoring, can be selected from proper subsets. Applications in technology are numerous. Some areas include energy consumption optimization, and network safe design. Figure 41 shows the geometric elements of an optimized water sky jump design found non intentionally. In Architecture apply all of the above, plus the aesthetic component. Many outprint suggestions reinforce the imagination when in a creative process. Figures 42 through 44 are examples of outputs for architects.

Painting, Sculpture, Music

All paintings and sculpture are subsets of numbers. They need special graph tools, the one that we are using here has limited capacity for this purpose. For painting is required one that plot pixels of tones of gray or color. Figure 53 shows the numeric matrix to print a primitive face. For sculpture one that plot solid 3D rendering, and for music one that writes Midi files. All of them work with the data files generated by the Polyn program. Figure 45 is an example of simple outprints, some interesting for textiles design. Figures 46, 47 and the Gallery are examples of simple outprints, try: Data: 1, 1, 1 Ranges: 1000 to 10000 Step 333, 10000 to 1000 Step –333, etc. column 5 vs 1 and column 9 vs 1. Due to the certainty of completeness of subsets of a set, the elements for music composition provide the full range for analysis of sound combinations. Automatic music composition has so far reached a dead end, because algorithms are mostly based in randomness that produce redundant sounds. The random approach is incredibly inefficient, but applying the theorem to sound is yet less workable to compose scores. The diskette accompanying this book contains generated melodies selected from random subsets, standard accompaniments were added to the tracks.

Economics, Financing

We can view these problems from three different perspectives:

1) The main function is $x + y + z = a$, or $3 + 5 + 7 = 15$

    and a first change produce $7 + 12 + 7 = 26$
    a second change produce $9 + 17 + 7 = 33$
    a third change produce $\quad 10 + 20 + 7 = 37$

then $37 - 15 = -22$, this is the maximum change from the original data. This is a dynamic process where changes in x, y and z are all valid in a differential of time dt.

2) The answer to max-min problems of the type:
   a) a, b, and c are benefits; x, y, and z are quantities that I can produce, what is the maximum benefit?
   b) a, b, and c are required investments; x, y, and z are quantities to produce, what is the minimum capital needed?
   c) A combination of both.

Example: Data 1, 2, 3. Constants: 6, 9, 12

$$a) \quad x \quad y \quad z \quad a \quad b \quad c$$
$$b) \quad a \quad b \quad c \quad x \quad y \quad z$$

1,	1, 2, 3,	6, 9, 12,	6,	18,	36,	60	Maximum Benefits
2,	1, 3, 2,	6, 9, 12,	6,	27,	24,	57	
3,	2, 1, 3,	6, 9, 12,	12,	9,	36,	57	
4,	2, 3, 1,	6, 9, 12,	12,	27,	12,	51	
5,	3, 1, 2,	6, 9,12,	18,	9,	24,	51	
6,	3, 2, 1,	6, 9, 12,	18,	18,	12,	48	Minimum Capital

3) These and other problems can be solved with Dynamic Programming (DP), Linear Programming, Games Theory, Arms Race Models for the Military, and so on. The DP approach can use a Markov Chain as a tool for solutions, it is a matrix of probabilities, such that having at time t_1 the probability of correlated events, we have the optimization of the outcome. This outcome is the initial conditions for another time t_2 where the probabilities of the matrix could be different.

Example (taken from the literature): The birth weights of English women and the birth weights of their daughters were split into three categories: low (below 6 pounds), average (between 6 and 8 pounds), and high (above 8 pounds). Among women whose own birth weights were low, 50 percent of the daughters had low birth weights, 45 percent had average weights, and 5 percent had high weights. Women with average birth weights had daughters with average weights half of the time, while the other half was split evenly between low and high categories. Women with high birth weights had female babies with high weights 40 percent of the rime, with low and average weights each occurring 30 percent of the time. This example can be considered as a Markov chain with three states (low, average, and high), corresponding to an experiment of choosing a woman at random and noting her birth weight. The translation matrix, easily derived from the verbal description, looks like this:

		Daugther Low	Average	High
	Low	0.50	0.45	0.05
P = Mother	Average	0.25	0.50	0.25
	High	0.30	0.30	0.40

Example: An electronics store has one item on special sale each day; it is either a television, a radio, or a stereo set. Stereos are never on sale two days in a row; if the store has a stereo as the special one dat, it is equally likely to have a TV or radio on special the next day. If the special one day is a TV or a radio, there is an even chance of continuing the item the next day. If the special item is changed from a TV or radio, only one-third of the rime will a stereo set be the special the next day. Consider this an example of Markov chain with states of TV, radio, and stereo. The transition matrix is:

		Tomorrow's Special		
		TV	Radio	Stereo
P = Today's Special	TV	0.50	0.33	0.16
	Radio	0.33	0.50	0.16
	Stereo	0.50	0.50	0

The solution of a problem can be found using matrices when applying a practical approach. In addition we have the option to investigate other subsets looking for suggestions on how to modify the problem to have a desirable solution. This modification can affect the data in the subset, the coefficients used, or both. If the coefficients (constant or range) represent probabilities and the data is kept unchanged, we have a simple approach for all cases, where the Markov Chain and matrices are hidden. How the work is done is presented as optimization of Stock Market Investment using probability coefficients:

Example: Stock Market Investment. A broker/individual wants to invest in a particular stock, the information he/she has is the following: (i) There are not more than four factors that lower the price, (ii) The probabilities of occurrence of these factors concurrently are certain, (iii) The probability coefficients are: 0.3, 0.1, 0.7 and 0.65 for any factor or between 0.1 and 0.7, (iv) These factors affect the total value of the stock in a 10%, 30%, 15% and 45% respectively, (v) These factors are valid at 2PM only, (vi) He/She has this information at 01:30 P.M., and (vi) He/She needs 5 minutes to order the operation.

At 01:30 He/She uses Data: 0.1, 0.3, 0.15, 0.45, and constants 0.3, 0.1, 0.7 and 0.65, with the following output:

	1	2	3	4	5	6	7	8	9	10
	Order	D1	D2	D3	D4	P1	P2	P3	P4	SUM
1	14	0.15	0.10	0.45	0.30	0.04	0.01	0.32	0.20	0.56
2	13	0.15	0.10	0.30	0.45	0.04	0.01	0.21	0.29	0.56
3	4	0.10	0.15	0.45	0.30	0.03	0.01	0.32	0.20	0.56
4	3	0.10	0.15	0.30	0.45	0.03	0.01	0.21	0.29	0.55

There are 3 maximums with 0.56. He/She take row 3 (originally row 4 before sorting the SUM column) because they want the minimum 0.03 for the first product as best option, then the selected subset is: $(0.1 * 0.3) + (0.15 * 0.1) + (0.45 * 0.7) + (0.30 * 0.65) = 0.56$. This maximum indicates that there are 56 percent chances that the price will be lower at 2 P.M.. At 01:55 He/She order to buy only a portion of the amount originally sought due to the low chance. With probability coefficients 0.7, 0.8, 0.9 and 0.95 the outcome is 90 percent likely that the price will drop.

Military, Intelligence

Tactics, strategy, arms race are mostly done with Game Theory. Intelligence has an approach similar to economic market behavior, but both can be solved with matrices and ultimately with probabilistic coefficients. A game is an attempt to represent and analyze mathematically some conflict situation in which the outcome depends on the choices made by the opponents. Game may be used to investigate problems in business, personal relationships, military maneuvers, and other areas involving decision-making, as extension of the military approach the same apply to business development, stock market, and economic/financial analysis. One particular kind of game for which the theory has been well developed is the matrix game. Suppose that, in a matrix game, $E(x, y)$ is the expectation, where x and y are mixed strategies for the two players. Then: $\max_x \min_y E(x, y) = \min_y \max_x E(x,y)$

By using a maximum strategy, one player, R, ensures that the expectation is at least as large as the left-hand side of the equation. Similarly, by using a minimax strategy, the other player, C, ensures that the expectation is less than or equal to the right-hand side of the equation. Such strategies may be called optimal strategies for R and C. Since, the two sides of the equation are equal, then if R and C use optimal strategies the expectation is equal to the common value, which is called the value of the game, e.g., consider the game given by the matrix:

$$\begin{vmatrix} 4 & 2 \\ 3 & 4 \end{vmatrix}$$

If $\underline{x} = (1/3, 2/3)$, it can be shown that $E(\underline{x}, y) >$ or $= 10/3$.

If $\underline{y} = (2/3, 1/3)$, then $E(x, \underline{y}) <$ or $= 10/3$ for all x. It follows that the value of the game is 10/3, and $\underline{x}$ and $\underline{y}$ are optimal strategy for the two players.

Example: Data: 1, 3, 7 (Sum = 11). Probabilistic coefficients (constants): 0.9, 0.7, and 0.2

	1	2	3	4	5	6	7	8	9	10
1	1	3	7	0.9	0.7	0.2	0.9	2.1	1.4	4.4
2	1	7	3	0.9	0.7	0.2	0.9	4.9	0.6	6.4
3	3	1	7	0.9	0.7	0.2	2.7	0.7	1.4	4.8
4	3	7	1	0.9	0.7	0.2	2.7	4.9	0.2	7.8
5	7	1	3	0.9	0.7	0.2	6.3	0.7	0.6	7.6
6	7	3	1	0.9	0.7	0.2	6.3	2.1	0.2	8.6

Best chances are given by max and min values, and other values speak of other possibilities in between. The Sum is not a product of probabilities, it is an optimizer as variable, the larger the best chance, certainty is given by 11. In this case maximum is 8.6 for the corresponding combination of possibilities or 78.18 % of certainty, and other possibilities can go as low as 4.4 or 40% of certainty.

Code and encryption are two areas where this methodology is well suited to be powerful by combining text and numbers, but no specifics are given here.

Prediction in Social Sciences

Let us take the example of the U.S. population. Using the logistic equation, and curve, it does not fit at the upper part, that seems to become flatter with large populations, see extrapolation A in Figure 48. It is necessary to change coefficients in the logistic equation based in previous data to get a presumable better approach as the extrapolated curve B in Figure 48. The same can be made by taking the previous data and using a subset. This result could be as good as or better than the change of coefficients in the logistic equation. Both are not simple extrapolation of existing data, but a more complex one. The function chosen to be linear in the logistic model, might be taken to be a polynomial of higher degree so that higher order effects of the size of population on the growth rate could be included. Additional factors might be attached to the differential equation to incorporate the concept that the rate of change of population is not only function of population but about what is happening over time as well, leading to probabilistic models of population growth, where Markov Chains can be applied. The estimates of the Census Bureau prove that the flat part of the logistic curve is located above the 2050 year mark. The logistic equation as a practical rule is not good enough for different time slices. Also this shows that the flat part of the curve, where the carrying capacity of the country, is larger than 394 million and will be found above the year 2050 mark. The logic behind the mathematical philosophy of both cases is a form of instinct presumption on how it will grow. None is exact, and the certainty is function of many probability coefficients. Assuming that probabilities to influence the outcome at the year 2020 from 2010, which is a 10-year segment, are: Immigration a factor of 0.02, Economics 0.05, and Birth/Death 0.08. Data: 298,000,000 - 298,000,000 - 298,000,000 Constants: 0.02, 0.05 and 0.08:

	1	2	3	4	5	6	7	8
		D1	D2	D3	P1	P2	P3	SUM
1	1	298000000	298000000	298000000	5960000	14900000	23840000	44700000

In 2020 we will have: 298,000,000 + 44,700,000 = 342,700,000. From 2020 through 2050 other runs will show that compounded probabilities will make them smaller and the closeness to certainty will vanish quickly. If data are 250, 300 and 350 millions due to probabilities at slice 2010, then at 2020 the minimum and maximum to add at the base data in 2010, once known, will be: 42 or 48 millions (and 43.5 or 46.5 millions in between)

U.S. Population Estimate:

	1	2	3
	Year	Actual Data	Logistic Output
1	1790	3929	3629
2	1800	5308	5336
3	1810	7240	7228
4	1820	9638	9757
5	1830	12866	13109
6	1840	17069	17506
7	1850	23192	23192
8	1860	31443	30412
9	1870	38558	39372
10	1880	50156	50177
11	1890	62948	62769
12	1900	75996	76870
13	1910	91972	91972
14	1920	105711	107395
15	1930	122775	122398
16	1940	131669	136318
17	1950	150697	148678
18	1960	179323	159230
19	1970	203185	167944

Next Output match the Census Estimate:

	1	2	3
	Year	Logistic Output	Census match
1	1790	4000	3000
2	1800	6000	5000
3	1810	8000	7000
4	1820	11000	10000
5	1830	14000	13000
6	1840	18000	17000
7	1850	24000	23000
8	1860	31000	30000
9	1870	41000	40000
10	1880	51000	50000
11	1890	64000	63000
12	1900	78000	77000
13	1910	93000	92000
14	1920	108000	107000
15	1930	123000	122000
16	1940	137000	136000
17	1950	150000	149000
18	1960	160000	170000
19	1970	168000	200000
20	1980	169000	230000
21	1996	-	265000
22	1997	-	267000
23	1998	-	270000
24	1999	-	272000

	1	2	3
	Year	Logistic Output	Census match
25	2000	-	274000
26	2001	-	277000
27	2002	-	279000
28	2003	-	281000
29	2004	-	284000
30	2005	-	286000
31	2006	-	288000
32	2007	-	290000
33	2008	-	293000
34	2009	-	295000
35	2010	-	298000
36	2015	-	310000
37	2020	-	322000
38	2025	-	335000
39	2030	-	347000
40	2035	-	358000
41	2040	-	370000
42	2045	-	381000
43	2050	-	394000

Formulation of Law in Natural Sciences

It is interesting to transcribe the following: "Systems with large numbers of variables may sometimes be optimized by a process called simulation, which involves trial and error with the actual system or a model. In simulation, the properties of the system or model are adjusted with a specific input or range of inputs to the system, and outputs or performance are measured until an optimum result is obtained" (**Merritt Frederick S., Standard Handbook of Civil Engineers, McGraw-Hill, New York, 1996, page 1.6**). We are using a set of polynomials as the shortest path to obtain a testable answer.
Let us elaborate three examples:

I.   For flood studies the USGS provides regression equations to know flow peak discharges of the form $Q = aA^b$ where Q is discharge in cfs and A is area in mi^2. This type of equations are known to have less precision when compared with actual measurements or using more parameters and hydraulic engineers use them as last resort. For instance the Shoal Creek in Metro Atlanta must use $Q = 1{,}010\ A^{0.584}$ where $A = 3.1\ mi^2$ and Q will be 1,956 cfs. But discharge is function of many parameters like s = soil saturation and tc = time of concentration in the catchment, actual h = water elevation in the channel, and the precipitation p that will increase the flow. Many times these information is not available and this is why the regression equation is used. In this particular case we know the Q in five cross sections in the area of interest at 900, 1400, 1800, 2500 and 3600 cfs, and s = 0.2, h = 10 ft, tc = 60 min. We need to know the discharge when p = 8 in.

It is necessary to stress that all hydrologic and hydraulic considerations are intrinsically included in the Q's values data. Looking for the Sums be equal to the Q's known, we use Data: 100, 200, 100, 500 Ranges 1 through 4 Step 1, and select from the output the following:

		Ranges				Data				Products			Sum
1	1	1	1	1	100	200	100	500	100	200	100	500	900
2	4	1	3	1	100	500	100	200	400	500	300	200	1400
3	3	2	3	1	100	500	100	200	300	1000	300	200	1800
4	2	3	4	2	100	500	100	200	200	1500	400	400	2500
5	4	4	4	4	100	500	100	200	400	2000	400	800	3600

The data is too large to work with, we will reduce the whole system in the following way:

	$a_1 \quad x^4$		$a_2 \quad x^3$		$a_3 \quad x^2$		$a_4 \quad x^1$		Sum = Q
Product	100		200		100		500		900
	1	100	1	200	1	100	1	500	
Reduction	0.16	625	1.6	125	4	25	100	x = 5	
Product	400		500		300		200		1400
	4	100	1	500	2	100	1	200	
Reduction	0.64	625	4	125	12	25	40	x = 5	
Product	300		1000		300		200		1800
	3	100	2	500	3	100	1	200	
Reduction	0.48	625	8	125	12	25	40	x = 5	
Product	200		1500		400		400		2500
	2	100	3	500	4	100	5	200	
Reduction	0.32	625	12	125	16	25	80	x = 5	
Product	400		2000		400		800		3600
	4	100	4	500	4	100	4	200	
Reduction	0.64	625	16	125	16	25	160	x = 5	

$$0.16\, x^4 + 1.6\, x^3 + 4\, x^2 + 100\, x = 900$$
$$\quad * \qquad\qquad\qquad *$$
$$0.64\, x^4 + 4\, x^3 + 12\, x^2 + 40\, x = 1400$$
$$0.48\, x^4 + 8\, x^3 + 12\, x^2 + 40\, x = 1800$$
$$\qquad\qquad * \qquad\qquad\qquad *$$
$$0.32\, x^4 + 12\, x^3 + 16\, x^2 + 80\, x = 2500$$
$$0.64\, x^4 + 16\, x^3 + 16\, x^2 + 160\, x = 3600$$

(*) are the interpolated locations for $a_1 = 0.2$, $a_2 = 10$, $a_3 = 8$, $a_4 = 60$

finally $\qquad\qquad\qquad\qquad\qquad$ **x = 5**

and the equation: $\qquad\qquad$ **$0.2\, x^4 + 10\, x^3 + 8\, x^2 + 60\, x = 1875$ cfs**

II.   Let us take the case of the force of gravitation, the well known formula of Newton is $F = G.M.m \,/\, d^2$ where $G = 6.7 \times 10^{-8}$ dyne cm^2/gr^2, and we will take the following data with its application:

M	m	d	d^2	$F = G M m / d^2$	$F = a_1 x^3 + 0.1 x^2 + 0.1 x$
10	1	100	10^4	67.0×10^{-12}	67.0×10^{-12}
30	3	446.84	199,668.87	30.2×10^{-12}	30.2×10^{-12}
50	5	1,000	10^6	16.6×10^{-12}	16.6×10^{-12}
100	10	10,000	10^8	0.6×10^{-12}	0.6×10^{-12}

Because we know the boundaries of our data, we will use Data: 1, 1, 1 Ranges: 0.1 to 70 Step 0.1 Running 2800 rows, sort decreasing the Sum column, and selecting, we obtain the following:

1	2	3	4	5
	Product 1	Product 2	Product 3	Sum
1	66.80	0.10	0.10	* 67.00
2	66.70	0.20	0.10	67.00
3	66.50	0.40	0.10	67.00
4	66.60	0.30	0.10	67.00
5	16.60	0.10	0.10	* 16.80
6	16.50	0.20	0.10	16.80
7	16.40	0.30	0.10	16.80
8	16.30	0.40	0.10	16.80
9	0.10	0.40	0.10	0.60
10	0.40	0.10	0.10	* 0.60
11	0.30	0.20	0.10	0.60
12	0.20	0.30	0.10	0.60

(*) these are the selected rows, and the equations valid in the ranges of $100 < d < 10,000$; $1 < m < 10$; and $10 < M < 100$, are:

$$66.8\, x^3 + 0.1\, x^2 + 0.1\, x = 67.00$$
$$16.6\, x^3 + 0.1\, x^2 + 0.1\, x = 16.80$$
$$0.4\, x^3 + 0.1\, x^2 + 0.1\, x = 0.60$$

then
$$x = 1$$

Now we want to know F for $M = 30$ and $m = 3$, we write: $F = 30\, x^3 + 0.1\, x^2 + 0.1\, x$, then

$$F = (6.7 / 10^8)\, ((30 \times 3) / d^2) \text{ or } d = 446.84 \text{ and } d^2 = 199,668.87$$

The general equation with F as function of M is:

$$F = a_1 x^3 + 0.1 x^2 + 0.1 x \quad \text{or} \quad F = (a_1 + 0.2) \times 10^{-12}$$

Therefore, within these boundaries we have a substitute of the Newton's Law of Universal Attraction. The result is inserted in italics in the table above.

III.   In this example we will see with a set of experimental data, and within its boundaries, a value obtained by regression based in mathematical considerations, and the same with inferential analysis. Therefore, in a natural phenomena or in human affairs, the most probable value is the one which is function of the states of the phenomena within its boundaries, where mathematics values are its descriptors. Data:

Dependent Variable	Independent Variable	Outcome
1	3	4
5	3	8
6	7	13
10	5	15
12	10	22
15	6	21

Use Data: 1, 1 Range 1 to 30 Step 1, Select the following:

	X	y	Outcome
1	1	3	4
2	5	3	8
3	6	7	13
4	10	5	15
5	12	10	22
6	15	6	21

With **x = 3** the transformed table is:

	a_1	x^2	a_2	x^1	Sum
Product		1		3	4
Reduction	0.11	9	1	x = 3	
Product		5		3	8
Reduction	0.55	9	1	x = 3	
Product		6		7	13
Reduction	0.66	9	2.33	x = 3	
Product		10		5	15
Reduction	1.11	9	1.33	x = 3	
Product		12		10	22
Reduction	1.33	9	3.33	x = 3	
Product		15		6	21
Reduction	1.66	9	2	x = 3	

$$0.11 \, x^2 + 1 \, x = 4$$
$$0.55 \, x^2 + 1 \, x = 8$$
$$0.66 \, x^2 + 2.33 \, x = 13$$

$$1.11\,x^2 + 1.33\,x = 15$$
$$\mathit{1.20\,x^2 + 2.06\,x = 17}$$
$$1.33\,x^2 + 3.33\,x = 22$$
$$1.66\,x^2 + 2\,x \quad = 21$$

If we have as data the outcome equal to 17 and abcissa equal to 10.8, then
$$10.8 + a_2\,3 = 17$$
or
$$(1.2 \times 3^2) + a_2\,3 = 17$$
then
$$a_2 = 2.06$$
and the equation is:
$$1.20\,x^2 + 2.06\,x = 17$$
that is inserted above in italics.

Point A in Figure 52 has the coordinates:
$$x = 1.2\,(9) = 10.8$$
$$y = 2.06\,(3) = 6.2$$

Compare with the linear regression equation $y = 0.3236\,x + 3.024$

These three cases are defined by a single equation. The most valuable benefit of a system of polynomials is the way by which they can describe very complex mathematics problems. In that sense, not a single polynomial describe a complex phenomena but a group of polynomials, much like a system of standard equations. The difference is that a system of standard equations is solved when are found common values of the variables that satisfy all equations, in other words, is the final or stable state of the phenomena they represent.

A system of polynomials with varying values of the variables describe intermediate states of an evolving phenomena and, as we see, contain much more information. Nonetheless, can be identified by few data, range, and selected descriptors of the phenomena. Any of the figures already seen and those in the Gallery, and what they may represent of a phenomena, can be defined with less than 20 numbers inserted in the Data - Range - Select scheme.

In Figures 54 and 55 are shown two graphs: the first how to build a system of polynomials to describe three time slices of a natural phenomena with error as small as the technique applied to obtain the data permit, and the second how to represent the outcome with the trend approach where at any time of the occurring phenomena are used past and future data with comparatively large errors and uncertainty.

PART IV

Guide for Engineering and Science Applications

We will analyze some cases that will permit to go into actual applications in engineering and science, they can be one of the following:

(i) Given an equation, find the equivalent system of polynomials
(ii) Given a system of polynomials, find the equivalent equation
(iii) Given data, find the system of polynomials
(iv) Given the set of polynomials with its explicit data, find the equation

(i)Given an equation, find the equivalent system of polynomials: Worksheet 1 shows one example, where the equation has one unknown and there is a straightforward solution. To start, transfer any given value to the right to form the Sum column of the polynomial. Worksheet 2 shows a case with two unknown, here there are more than one polynomial as solution for the unknowns, but the basic is the one that have the unknown values in both summands.

(ii) Given a system of polynomials, find the equivalent equation: We can start with the basic in Worksheet 2 and proceed inversely, that is $1 \, x^2 + 3 \, x = 4$ then $(2/d) + (6/d) = 4$ then $d = 2$, and the equation is $(2 + 6) \, / \, d = 4$

e.g., the set

$$0.11 \, x^2 + 1 \, x^1 = 4$$
$$0.55 \, x^2 + 1 \, x^1 = 8$$

for x =3, is equivalent to the system of equations:

$$(1 + 3) \, / \, d = 4$$
$$(5 + 3) \, / \, d = 8$$

(iii) Given data, find the set of polynomials: Worksheets 3 and 4 are examples of this case. In particular, worksheet 4 shows data to assigned variables.

(iv) Given the set of polynomials with its explicit data, find the equation: Worksheet 5 shows that the polynomials does not produce an equation because the use of raw data for the variables under the boundaries of the range. Therefore, the polynomial in itself is the final equation, or a set of polynomials is in itself a system of equations. Raw data can produce a set of polynomials without passing by the equation step. Worksheet 6 shows the case where the set of polynomials takes data from a phenomena described by complex equations and form the set of polynomials to describe the phenomena. Using a curve trend out of the polynomials it is possible to obtain a power equation, exponential equation, or other as desired.

Regarding precision:

Raw data is discrete and we are able to use increments as small as we want when preparing the vertical data vectors in the set of polynomials. Because the set of polynomials define a phenomena fully with as many variables and boundaries as needed then graphs are not important, but can be constructed from the set, e.g., the vectors X = 1, 2, 3, 4 and Y = 10, 20, 30, 40 represent a 45 degree line in Cartesian coordinates scaling X vs. Y by 10. We can find any value of X between 1 and 4, and Y between 10 and 40, using x = 1, the order of the polynomials is 2, and the range 10 to 40 step 10 for Y, to get that line:

$$1\,x^2 + 10\,x^1 = 11$$
$$2\,x^2 + 20\,x^1 = 22$$
$$3\,x^2 + 30\,x^1 = 33$$
$$4\,x^2 + 40\,x^1 = 44$$

How to obtain a set of polynomials:

The purpose is to find a set of polynomials that represent the phenomena. The job is to find the set even if it is defined by an extensive mathematical system or a large data log. We can take three cases for review:

Case 1: Simple formula or equation. With one unknown is deterministic as seen in worksheet 1. With two unknowns has no solution as seen in Worksheet 2, but we can approximate an acceptable solution within the range of **x** or **d** in the example.

Case 2: Solution by matrix. It is a polynomial or set of polynomials according the rules of the matrices, or operations of the determinants.

Case 3: Data from graphs. Import data from a graph into the set of polynomials. It is a determinist situation, data from the graph can be considered raw data or data obtained by using an equation.

In general, it s necessary to identify the variables that will form the vertical data vectors in the set of polynomials. Data for the variables, of any origin (experimental, from natural laws, from graphs), and the known or desired outcome will form the polynomials and Sums in the set.

After obtaining a set, sort the values of the variables or the sums in the spreadsheet and select those desired. After that, start to erase all polynomials which have variables out of the boundaries selected previously. At that point the set is complete, and it is possible to interpolate, graph, or derive any conclusion. If there are more unknown than polynomials, simply do not delete those previously thought to be redundant.

WORKSHEET 1

Part A

				Data	Polyn Value
a_1				6	4
a_2				2	4

$$\frac{a+b}{c} = d \qquad\qquad \frac{2+6}{4} = d$$

$$\frac{2+6}{d} = 4 \qquad\qquad \frac{2}{d} + \frac{6}{d} = 4$$

Part B

Selected Set of Polynomials. Data 0.5 , 0.5 - Ranges: 1 thru 6 Step 1

	Ranges						Data						Products					Sum
1				2	6				0.5	0.5				1	3			4
2				4	4				0.5	0.5				2	2			4
3				3	5				0.5	0.5				1.5	2.5			4

Reduction with x = 1

		a_6	a_5	a_4	a_3	a_2	a_1
1	$x^1 = 1$					1	3
2	$x^2 = 1$ Product / x^i						

Part C

Math Model

$$1\ x^2 + 3\ x^1 = 4$$

Comment: This is a simple equivalence of

$$1\ x^2 + 3\ x = 4 \qquad \text{for } x = 1 \text{ with}$$
$$(2/d) + (6/d) = 4 \qquad \text{where } d = 2$$

WORKSHEET 2

Part A

	Data	Polyn Value
a_1	2	4

$$\frac{a+b}{c} = d \qquad \frac{2+x}{4} = d$$

$$\frac{2+x}{d} = 4 \qquad \frac{2}{d} + \frac{x}{d} = 4$$

Part B

Selected Set of Polynomials. Data: 0.5 , 0.5 - Ranges: 1 thru 6 Step 1

	Ranges						Data					Products				Sum
1					4	4				0.5	0.5				2	2	4
2					5	3				0.5	0.5				2.5	1.5	4
3					5	3				0.5	0.5				2.5	1.5	4
4					3	5				0.5	0.5				1.5	2.5	4
5					3	5				0.5	0.5				1.5	2.5	4
6					4	4				0.5	0.5				2	2	4

Reduction with x = 1

		a_6	a_5	a_4	a_3	a_2	a_1
1	$x^1 = 1$ $x^2 = 1$					2	2
2	$x^3 =$ $x^4 =$					2.5	1.5
3	$x^5 =$ $x^6 =$					2.5	1.5
4						1.5	2.5
5	Product / x^i					1.5	2.5
6						2	2

Part C
Math Model

$$2.0\ x^2 + 2.0\ x^1 = 4$$
$$2.5\ x^2 + 1.5\ x^1 = 4$$
$$1.5\ x^2 + 2.5\ x^1 = 4$$
$$2.0\ x^2 + 2.0\ x^1 = 4$$

Comment: $2x^2 + 2x = 4$ for x = 1   equivalent to $(2/d) + (x/d) = 4$
It is possible to interpolate: $2x^2 + 1.7x = 3.7$, or extrapolate using more polynomials.

Part B

		Data	Polyn Value
Examples of Interpolation	a_1		4
	a_2		8
See Figures 54 and 55	a_3		13
	a_4		15
	a_5		22
	a_6		21

Part B

Selected Set of Polynomials. Data: 1 , 1 - Ranges: 1 thru 30 Step 1

	Ranges					Data					Products					Sum
1				1	1				1	3				1	3	4
2				1	1				5	3				5	3	8
3				1	1				6	7				6	7	13
4				1	1				10	5				10	5	15
5				1	1				12	10				12	10	22
6				1	1				15	6				15	6	21

Reduction with x = 3

		a_6	a_5	a_4	a_3	a_2	a_1
1	$x^1 = 3$ $x^2 = 9$					0.11	1
2	$x^3 =$ $x^4 =$					0.55	1
3	$x^5 =$ $x^6 =$					0.66	2.33
4						1.11	1.33
5	Product / x^1					1.33	3.33
6						1.66	2

Part C
Math Model

$$0.11\ x^2 + 1.00\ x^1 = 4$$
$$0.55\ x^2 + 1.00\ x^1 = 8$$
$$0.66\ x^2 + 2.33\ x^1 = 13$$
$$1.11\ x^2 + 1.33\ x^1 = 15$$
$$1.33\ x^2 + 3.33\ x^1 = 22$$
$$1.66\ x^2 + 2.00\ x^1 = 21$$

Comment: e.g., $1.20\ x^2 + 2.06\ x = 17$ out of a set of six polynomials

Part A

		Data	Polyn Value
Peak Flow Discharge: Shoal Creek, Metro Atlanta	A_1	s = 0.2	900
	A_2	h = 10	1400
	A_3	tc = 60	1800
Interpolation with multiple variables.	A_4	p = 8	2500
	A_5		3600

Part B

Selected Set of Polynomials. Data: 100 , 200 , 100 , 500 - Ranges: 1 thru 4 Step 1

	Ranges				Data				Products				Sum
1	1	1	1	1	100	200	100	500	100	100	200	500	900
2	4	1	3	1	100	500	100	200	400	500	300	200	1400
3	3	2	3	1	100	500	100	200	300	1000	300	200	1800
4	2	3	4	2	100	500	100	200	200	1500	400	400	2500
5	4	4	4	4	100	500	100	200	400	2000	400	800	3600

Reduction with x = 5

		a_6	a_5	a_4 **s**	a_3 **h**	a_2 **p**	a_1 **tc**
1	$x^1 = 5$			0.16	1.60	4	100
2	$x^2 = 25$ $x^3 = 125$			0.64	4	12	40
3	$x^4 = 625$			0.48	8	12	40
4	$x^5 = 3125$ $x^6 = 15625$			0.32	12	16	80
5	Product / x^i			0.64	16	16	160

Part C
Math Model

$$0.16\,x^4 + 1.6\,x^3 + 4\,x^2 + 100\,x = 900$$
$$0.64\,x^4 + 4\,x^3 + 12\,x^2 + 40\,x = 1400$$
$$0.48\,x^4 + 8\,x^3 + 12\,x^2 + 40\,x = 1800$$
$$0.32\,x^4 + 12\,x^3 + 16\,x^2 + 80\,x = 2500$$
$$0.64\,x^4 + 16\,x^3 + 16\,x^2 + 160\,x = 3600$$

Comments: Example of multiple column interpolation:
$0.2\,x^4 + 10\,x^3 + 8\,x^2 + 60\,x^1 = 125 + 1250 + 200 + 300 = 1{,}875$
Note how the selection of a proper x value make the range of the variables fit
out of a set of five polynomials.

Part A

		Data	Polyn Value
Law of Gravitation $F = G\, Mm / d^2$	a_1	0.1	67.00
	a_2	m = 0.1	16.80
	a_3	M = 30	0.60

Part B

Selected Set of Polynomials. Data: 1 , 1 , 1 - Ranges: 1 thru 70 Step 0.1

	Ranges			Data			Products			Sum
1	1	1	1	66.8	0.1	0.1	66.8	0.1	0.1	67.00
2	1	1	1	16.6	0.1	0.1	16.6	0.1	0.1	16.80
3	1	1	1	0.4	0.1	0.1	0.4	0.1	0.1	0.60

Reduction with x = 1

		a_6	a_5	a_4	a_3	a_2	a_1
1	$x^1 = 1$				66.80	0.1	0.1
2	$x^2 = 1$ $x^3 = 1$				16.60	0.1	0.1
3	Product / x^i				0.40	0.1	0.1

Part C

Math Model

$$66.8\ x^3 + 0.1\ x^2 + 0.1\ x^1 = 67.00$$
$$16.6\ x^3 + 0.1\ x^2 + 0.1\ x^1 = 16.80$$
$$0.40\ x^3 + 0.1\ x^2 + 0.1\ x^1 = 0.60$$

Comments:

The system of polynomials is valid for the established following ranges:

$$100 < d < 10,000$$
$$1 < m < 10$$
$$10 < M < 100$$

where $F = G\ Mm/d^2$ is   equivalent   to $F = a_3\ x^3 + a_2\ x^2 + a_1\ x^1$

Part A

		Data	Polyn Value
Hydraulic Application: **Moody Diagram** **(Churchill Formula)** $f = 8\left[\,(8/Re)^{12} + (A+B)^{-3/2}\,\right]^{1/12}$ $A = \left[2.457\ \mathrm{Ln}\left(1/\left((7/Re)^{0.9} + 0.27\,(e/D)\right)\right)\right]^{16}$ $B = (37530/Re)^{16}$	a_1	Re: 20K to 100K	20,000
	a_2	e/D: 0.0001 to 0.006	40,000
	a_3		100,000

Part B

Selected Set of Polynomials. Data: 1 , 1 , 1   Ranges: 0.0001 thru 100,000 Step 0.001

	Ranges			Data			Products			Sum
1	1	1	1	0.006	0.036	19,999.958	0.006	0.036	19,999.958	20,000
2	1	1	1	0.002	0.027	39,999.971	0.002	0.027	39,999.971	40,000
3	1	1	1	0.0001	0.0185	99,999.9814	0.0001	0.0185	99,999.9814	100,000

Reduction with x = 1

		a_6	a_5	a_4	a_3 e/D	a_2 f	a_1 Re
1	$x^1 = 1$				0.006	0.036	19,999.958
2	$x^2 = 1$ $x^3 = 1$				0.002	0.027	39,999.971
3	Product / x^i				0.0001	0.0185	99,999.9814

Part C

Math Model

$$0.0060\ x^3 + 0.0360\ x^2 + 19{,}999.958\ x^1 = 20{,}000$$
$$0.0020\ x^3 + 0.0270\ x^2 + 39{,}999.971\ x^1 = 40{,}000$$
$$0.0001\ x^3 + 0.0185\ x^2 + 99{,}999.981\ x^1 = 100{,}000$$

Comments:

The intention is to define the curve with polynomials, such that f = f (Re, e, D).

Boundaries are taken properly for the different range of the variables. Example of interpolation:

   what is the Re value for e/D = 0.004 and f = 0.0315?, answer: Re = 30,000

The set of polynomials is simpler than Moody/Churchill, and make it comparable with other sets of similar phenomena. The set of polynomials replace Moody in the given range with x = 1. The whole Moody Graph can be replaced by an appropriate set of polynomials using a proper x-value.

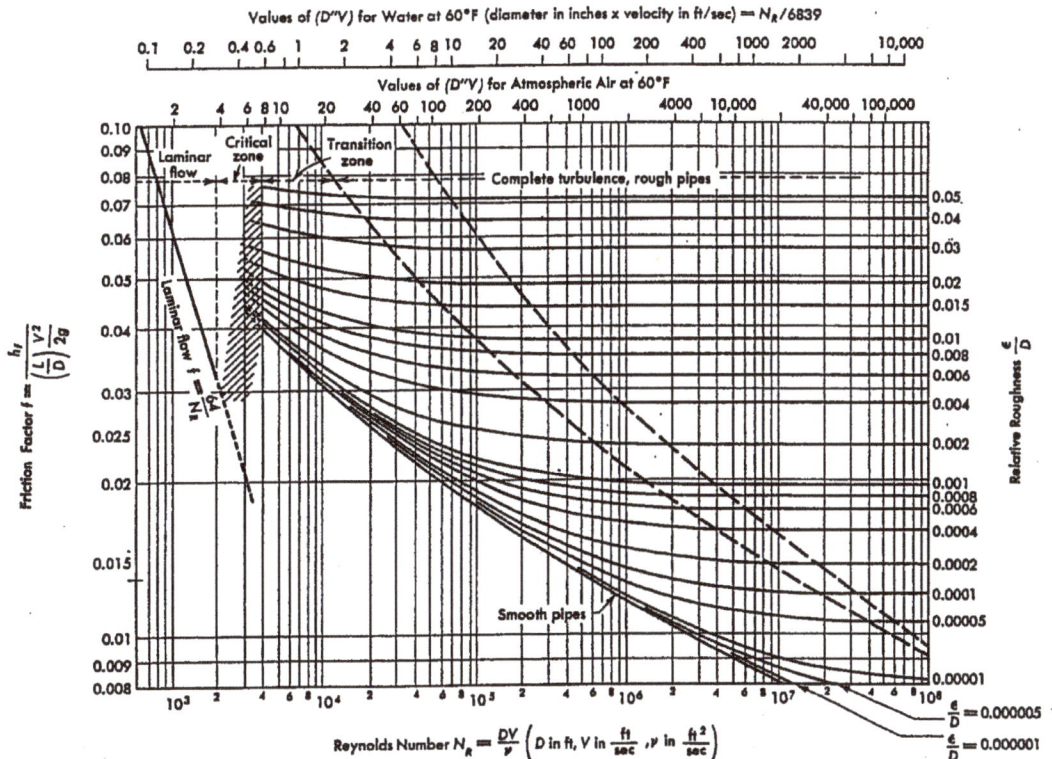

Moody Diagram
(from Applied Hydraulics in Engineering, Henry M. Morris and James M. Wiggert, John Wiley, New York, 1971)

Scientific Approach:

Case 1: Transform an analytic expression into a set of polynomials.

Let us take the formula $V_0 = P_1 V_1 / P_0$
And we know the values:

P_1	V_1	P_0	V_0
20	12	10	24
21.9	14	12	25.5

Running Polyn with Data: 1, 1, 1 and Ranges 0.1 to 22 step 0.1, 0.1 to 14 step 0.1, and 0.1 to 12 step 0.1, and selecting (where x = 1):

$$20.0\ x^3 + 3.9\ x^2 + 0.1\ x = 24.0$$
$$\mathbf{21.0\ x^3 + 3.7\ x^2 + 0.5\ x = 25.2}$$
$$21.9\ x^3 + 3.5\ x^2 + 0.1\ x = 25.5$$

The interpolated polynomial was obtained by selecting 25.2 in the data run when looking for the functional variable = 21.0, and 3.7 and 0.5 were selected, then

$21.0\ x^3 + 3.7\ x^2 + 0.5\ x = 25.2$ is equivalent to $V_0 = P_1 V_1 / P_0$ or $(21 \times 12) / 10 = 25.2$

Case 2: Transform a set of polynomials into an analytic expression.

First we need to know the raw data from the field or the lab, it is an obvious problem, simply each variable form a vertical data vector, where the variables are the coefficients of x^i and the sums are obtained by adding the products.

e.g., vector 1: 1, 3, 4, 6, 8, 10, 12, 14
      vector 2: 3, 5, 7, 9, 12, 15, 17, 18

for x = 2 the set of polynomials is:

$$1\ x^2 + 3\ x = 10$$
$$3\ x^2 + 5\ x = 22$$
$$4\ x^2 + 7\ x = 30$$
$$6\ x^2 + 9\ x = 42$$
$$8\ x^2 + 12\ x = 56$$
$$10\ x^2 + 15\ x = 70$$
$$12\ x^2 + 17\ x = 82$$
$$14\ x^2 + 18\ x = 92$$

which represent and substitute a known or <u>unknown analytic expression</u>. Must be a correspondence between a set of polynomials and an analytic expression, if no correspondence is found then a coefficient(s) need to be added to the analytic expression (e.g., the Constant of Gravitation in its formula). If a single value of the vector is the functional variable, then its polynomial is the analytic expression (or formula). The data above can be written analytically as $c = e^{kb}$

The table below is an example of substitution of an approximated system by an exact one, data is X, Y and result is Z.

MULTIPLE LINEAR REGRESSION – TWO INDEPENDENT VARIABLES
$$Z = A + B*X + C*Y$$
$$Z = 10406.05 + (-28.12477 * X) + (58.33419 * Y)$$
$$R^2 = 0.9959109$$

I	X	Y	Z	XCALC	ZCALC - Z
1	410	150	7500	7625.021	125.0205
2	370	150	8750	8750.012	1.171875E-02
3	330	150	10000	9875.003	-124.9971
4	410	140	7040	7041.679	1.678711
5	370	140	8170	8166.67	-3.330078
6	330	140	8290	9291.661	1.661133
7	410	130	6460	6458.337	-1.663086
8	370	130	7580	7583.328	3.328125
9	330	130	8710	8708.319	-1.680664
10	410	120	6000	5874.995	-125.0049
11	370	120	7000	6999.987	-1.367188E-02
12	330	120	8000	8124.978	124.9775

Running POLYN with the following boundaries (1):

X = 330 to 410 step 10
Y = 120 to 150 step 10
Z = 6000 to 10000 step 10

For x = 1, and selecting the pair X, Y as given, the exact Z's are obtained by sorting any column. Therefore, (1) substitute the table above, and both represent an unknown analytical expression, which is easy to find under the standard mathematical procedures.

Engineering Approach:

In engineering are used equations, formulas, nomographs, computer programs, etc. We will show how these are transformed, and how an engineering textbook will look like with this general methodology. Exhibits A and B are a standard engineering page presentation and its nomographs to facilitate computations. To have data, we will bound the extend of the nomograph with the following set of polynomials, where the data vectors are: a_1 is Manning's n, a_2 is Diameter, a_3 is Flow, and a_4 is Velocity. The run was made by using POLYN with a_3 (Flow) as the functional or leading variable, a set of polynomials was obtained, where the boundaries were:

$$v = 1 \text{ to } 15 \text{ step } 0.1$$
$$\text{Flow} = 1 \text{ to } 1000 \text{ step } 1$$
$$\text{Dia} = 6 \text{ to } 120 \text{ step } 1$$
$$n = 0.0001 \text{ to } 0.1 \text{ step } 0.0001$$

for n = 0.015

$$15.0\ x^4 + 1000\ x^3 + 108\ x^2 + 0.0800\ x^1 = 1{,}123.08$$
$$12.7\ x^4 + 1000\ x^3 + 120\ x^2 + 0.0051\ x^1 = 1{,}132.75$$
$$\mathbf{9.65\ x^4 + 750\ x^3 + 120\ x^2 + 0.0030\ x^1 = 879.65}$$
$$6.6\ x^4 + 500\ x^3 + 120\ x^2 + 0.0014\ x^1 = 626.61$$
$$15.0\ x^4 + 500\ x^3 + 78\ x^2 + 0.0130\ x^1 = 593.02$$
$$15.0\ x^4 + 250\ x^3 + 54\ x^2 + 0.0200\ x^1 = 319.02$$
$$3.2\ x^4 + 250\ x^3 + 120\ x^2 + 0.0032\ x^1 = 373.20$$
$$15.0\ x^4 + 100\ x^3 + 34.5\ x^2 + 0.0380\ x^1 = 149.53$$
$$x = 1$$

Therefore, the expressions above can substitute those two engineering pages. If the unknown is the velocity for flow equals to 750, averaging sums and functional variable, will give as interpolation (see line AB in Exhibit 2), then **Velocity = 879.65 – 750 – 120 – 0.0030 = 9.65**
The set can become large and must be handled by a computer. To narrow the set around the solution, a run(s) with closer bounds is done, many examples in the previous pages shows how to become proficient in this technique. A text book may have data and ranges instead of equations, formulas or a printed set of polynomials. The advantages are the following:

- They avoid any analytic expression; it is possible to work with the desired precision.
- The set of polynomials involves only sums and products.
- Texts book or technical manuals can have this general and only method for computations, leaving mathematical analysis for theory development.

Formulas for Gravity and Pressure Flow

The most widely used formula for determining the hydraulic capacity of storm drain pipes for gravity and pressure flows is the Manning Formula and it is expressed by the following equation:

$$V = [1.486 \, R^{2/3} S^{1/2}]/n \tag{3.13}$$

Where:
$V =$ mean velocity of flow (ft/s)
$R =$ the hydraulic radius (ft) - defined as the area of flow divided by the wetted flow surface or wetted perimeter (A/WP)
$S =$ the slope of hydraulic grade line (ft/ft)
$n =$ Manning's roughness coefficient

In terms of discharge, the above formula becomes:

$$Q = [1.486 \, AR^{2/3} S^{1/2}]/n \tag{3.14}$$

Where:
$Q =$ rate of flow (cfs)
$A =$ cross sectional area of flow (ft^2)

For pipes flowing full, the above equations become:

$$V = [0.590 \, D^{2/3} S^{1/2}]/n \tag{3.15}$$
$$Q = [0.463 \, D^{8/3} S^{1/2}]/n \tag{3.16}$$

Where: $D =$ diameter of pipe (ft)

The Manning's equation can be written to determine friction losses for storm drain pipes as:

$$H_f = [2.87 \, n^2 V^2 L]/[S^{4/3}] \tag{3.17}$$
$$H_f = [29 \, n^2 V^2 L]/[(R^{4/3})(2g)] \tag{3.18}$$

Where:
$H_f =$ total head loss due to friction (ft)
$n =$ Manning's roughness coefficient
$D =$ diameter of pipe (ft)
$L =$ length of pipe (ft)
$V =$ mean velocity (ft/s)
$R =$ hydraulic radius (ft)
$g =$ acceleration of gravity $= 32.2$ ft/sec^2

3.7.4 Nomographs And Table

The nomograph solution of Manning's formula for full flow in circular storm drain pipes is shown on Figures 3-12, 3-13, and 3-14. Figure 3-15 has been provided to solve the Manning's equation for part full flow in storm drains.

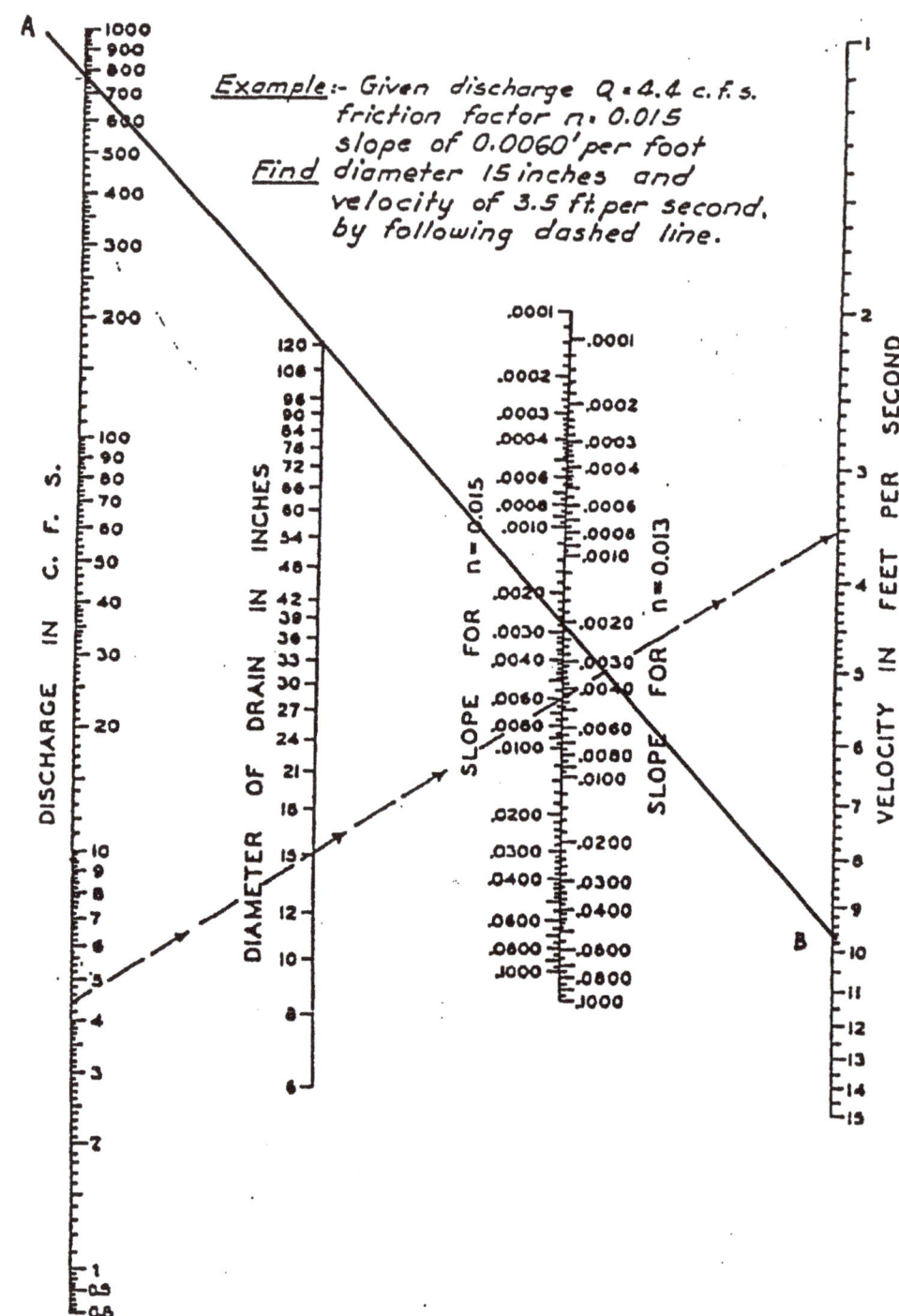

Figure 3-13 Nomograph For Computing Required Size Of Circular Drain, Flowing Full
n = 0.013 or 0.015

We can summarize the whole process in the following diagram:

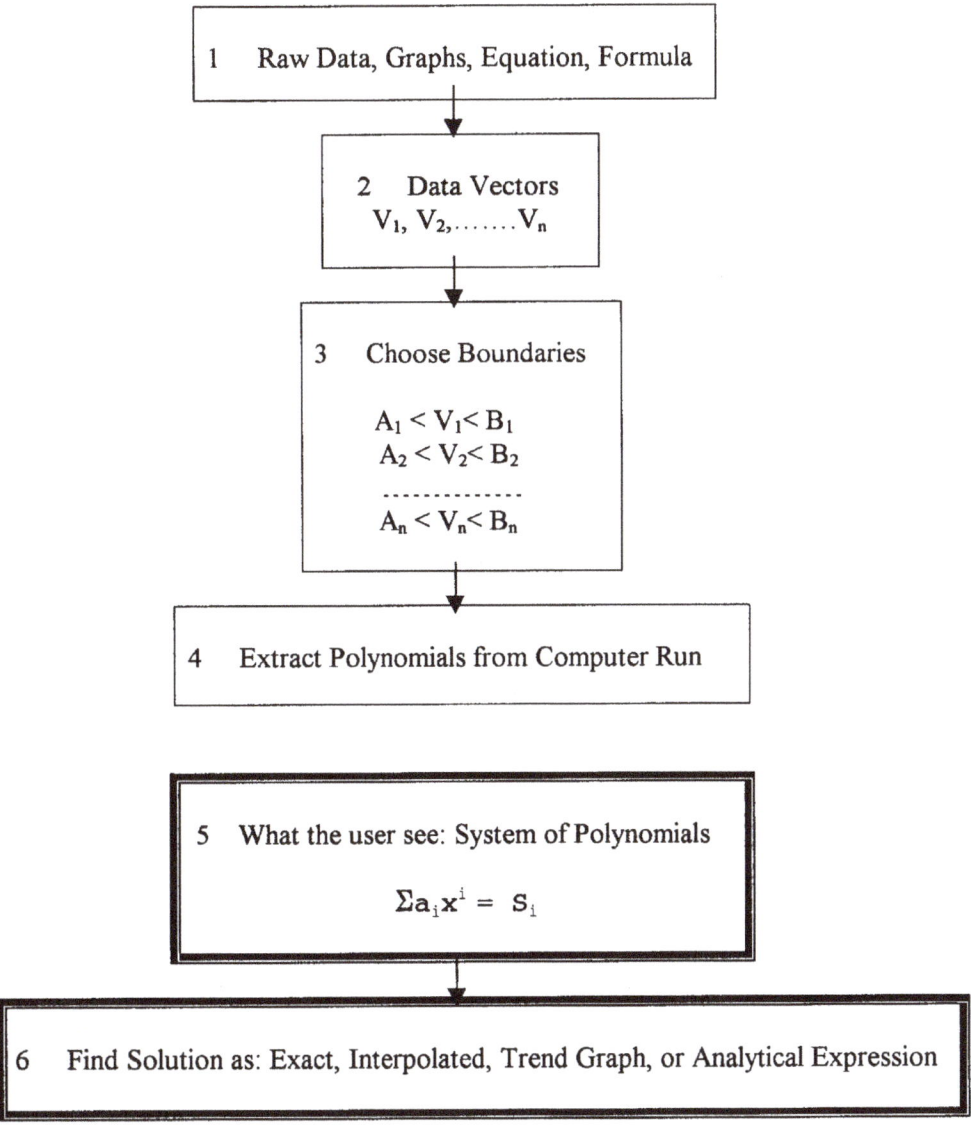

General comments follows:

1.   The system of polynomials reduces a physical law to sums, the simplest arithmetic operation.

2.   Do not confound a data table with a system of polynomials, the former is inert, the latter replaces a physical law, known or unknown, and have properties.

3. The properties of a system of polynomials are the minimum and sufficient for every mathematical operation.

4. The most important property of a system of polynomials is the one that makes all problems deterministic with an exact solution, by permutation of the vertical data vector within bounded limits and precision as desired.

5. It is feasible to represent all and any physical law, known or unknown, by using systems of polynomials.

To be considered an AI result, we introduced to some degree our reasoning when selecting one or more polynomials. Therefore, the whole process is not a full AI result but a practical one, in this matter we must recall Part I when we applied some degree of reasoning to select text logically perfect to our mind, or when we inferred conclusions from some segment of all probabilities in Part II.

The next three pages show the three-part worksheet suggested to work out the set of polynomials, where the first table of Part B is obtained by running the software POLYN. The second table of Part B and Part C can be obtained by using the program MODEL included in the diskette, it runs in DOS or WINDOWS. A printout sample is shown below:

SET OF POLYNOMIALS AS:

$$a_1.x^8 + a_2.x^7 + a_3.x^6 + a_4.x^5 + a_5.x^4 + a_6.x^3 + a_7.x^2 + a_8.x^1 = C$$

```
-------------------------------- DATA --------------------------------
   90      80      70      60      50      40      30      20   =  440
   88      78      68      58      48      38      28      18   =  424
   86      76      66      56      46      36      26      16   =  408
   84      74      64      54      44      34      24      14   =  392
   82      72      62      52      42      32      22      12   =  376
   80      70      60      50      40      30      20      10   =  360
   78      68      58      48      38      28      18       8   =  344
   76      66      56      46      36      26      16       6   =  328
----------------------- REDUCED   COEFFICIENTS -----------------------

 .35156  .625    1.09375
                         1.875   3.125   5       7.5     10   = 439.999
 .34375  .60937  1.0625  1.8125  3       4.75    7       9    = 423.999
 .33593  .59375  1.03125
                         1.75    2.875   4.5     6.5     8    = 407.998
 .32812  .57812  1       1.6875  2.75    4.25    6       7    = 391.998
 .32031  .5625   .96875  1.625   2.625   4       5.5     6    = 375.999
 .3125   .54687  .9375   1.5625  2.5     3.75    5       5    = 359.999
 .30468  .53125  .90625  1.5     2.375   3.5     4.5     4    = 343.998
 .29687  .51562  .875    1.4375  2.25    3.25    4       3    = 327.998

                              X = 2
----------------------------------------------------------------------
```

WORKSHEET

PART A

Equation or Formula	Raw Data	Data Guide		Polyn Value
		Lower Bound	Upper Bound	
	a_1			
	a_2			
	a_3			
	a_4			
	a_5			
	a_6			
	a_7			
	a_8			

Polyn Search

	a_8	a_7	a_6	a_5	a_4	a_3	a_2	a_1
Factor	1	1	1	1	1	1	1	1
Lower Bound								
Upper Bound								
Precision								

PART B
Selected Set of Polynomials

	Ranges	Data	Products	Polyn Value
1				
2				
3				
4				
5				
6				
7				
8				

Reduction with x =

	a_8	a_7	a_6	a_5	a_4	a_3	a_2	a_1
1	$x^1 =$; Product / $x^1 =$							
2	$x^2 =$; Product / $x^2 =$							
3	$x^3 =$; Product / $x^3 =$							
4	$x^4 =$; Product / $x^4 =$							
5	$x^5 =$; Product / $x^5 =$							
6	$x^6 =$; Product / $x^6 =$							
7	$x^7 =$; Product / $x^7 =$							
8	$x^8 =$; Product / $x^8 =$							

PART C

Math Model:

$$x^8 + x^7 + x^6 + x^5 + x^4 + x^3 + x^2 + x^1 =$$

$$x^8 + x^7 + x^6 + x^5 + x^4 + x^3 + x^2 + x^1 =$$

$$x^8 + x^7 + x^6 + x^5 + x^4 + x^3 + x^2 + x^1 =$$

$$x^8 + x^7 + x^6 + x^5 + x^4 + x^3 + x^2 + x^1 =$$

$$x^8 + x^7 + x^6 + x^5 + x^4 + x^3 + x^2 + x^1 =$$

$$x^8 + x^7 + x^6 + x^5 + x^4 + x^3 + x^2 + x^1 =$$

$$x^8 + x^7 + x^6 + x^5 + x^4 + x^3 + x^2 + x^1 =$$

$$x^8 + x^7 + x^6 + x^5 + x^4 + x^3 + x^2 + x^1 =$$

Engineering Examples

Case 1: Given an equation, find the equivalent system of polynomials

VERTICAL STRESS DUE TO A POINT LOAD

Analytic expression: Boussinesq (1883) solved the problem of stresses produced at any point in a homogeneous, elastic, and isotropic medium as the result of a point load applied on the surface of an infinitely large space.

The vertical normal stress is: $\Delta p = (P / z^2) I$, where P is the load in lb, z is the depth below ground surface, and I is function of r/z where r is distance from boundary to point of application of the load.

Data: We will use: r/z and I to obtain Δp.

Boundaries: r/z For 0.1 to 20 Step 0.1
              I For 1 to 0.001 Step 0.001

System of polynomials: Total rows will be n!(a)(b) = 2(200)(400) = 400,000
The set is too large and will be contained in an electronic file. Here we select a segment that includes the solution (in bold) to the following problem: "Consider a point load P = 1000 lb. Find the maximum vertical stress increase with depth due to the point load below ground surface with x = 3 ft and y = 4 ft (then r = 5 ft)." The leading variable will be r/z, then the format of the polynomials is $(r/z) x^2 + (I) x = \Delta p$, and the eight row set with x = 1 is:

$$2.50 \, x^2 + 0.003 \, x = 0.85$$

$$1.25 \, x^2 + 0.042 \, x = 2.65$$

$$\mathbf{0.83 \, x^2 + 0.129 \, x = 3.60 \ \ lb/ft^2}$$

$$0.50 \, x^2 + 0.273 \, x = 2.73$$

$$0.33 \, x^2 + 0.371 \, x = 1.65$$

$$0.25 \, x^2 + 0.410 \, x = 1.03$$

Case 2: Given very well known data, find the system of polynomials

FROM THE LABORATORY

Data: Vector A: 2, 4, 6, 7, 8, 5, 6, 7
       Vector B: 17, 23, 19, 18, 14, 18, 20, 16
       Vector C: 3.3, 4.7, 5.4, 6.6, 4.5, 5.2, 3.9, 6.1

Boundaries: A From 2 To 7 Step 1
             B From 14 To 23 Step 1
             C From 3.3 To 6.6 Step 0.1
Total rows: $n!$. a . b. c = 40320 (7) (9) (33) = 83,825,280

Analytic expression: None

System of polynomials: Because the size of the file, we will narrow to a segment of our interest, namely for the variable of the vector A, keeping constants B at 15, and C at 4.
Therefore the quantity of rows is 3! (7) (1) (1) = 42. We are looking for a maximum, it is given by the sum of the first polynomial shown below for x = 1:

$$7 x^3 + 15 x^2 + 4 x = 26$$
$$6 x^3 + 15 x^2 + 4 x = 25$$
$$5 x^3 + 15 x^2 + 4 x = 24$$
$$4 x^3 + 15 x^2 + 4 x = 23$$
$$3 x^3 + 15 x^2 + 4 x = 22$$
$$2 x^3 + 15 x^2 + 4 x = 21$$

With x = 2:

$$0.875 x^3 + 3.75 x^2 + 2 x = 26$$
$$0.750 x^3 + 3.75 x^2 + 2 x = 25$$
$$0.625 x^3 + 3.75 x^2 + 2 x = 24$$
$$0.500 x^3 + 3.75 x^2 + 2 x = 23$$
$$0.375 x^3 + 3.75 x^2 + 2 x = 22$$
$$0.250 x^3 + 3.75 x^2 + 2 x = 21$$

Scientific Examples

Case 1: Transform an analytic expression into a set of polynomials

GREENHOUSE EFFECT

Analytic expression: A simple climate model represents the Earth as a lumped mass having a uniform temperature and uniformly distributed thermal properties. The change in temperature of such a lumped mass is governed by the following equation:

$C(dT'/dt) + \lambda T' = Q'$, where C is the Earth's heat capacity, T' is the temperature change due to a change in global heating, Q' is the change in global heating due to changes in the amount of atmospheric CO_2.

The quantity $\lambda T'$ is a feedback term. It represents the physical processes that lead to additional cooling or heating of the atmosphere as the temperature increases. Units of λ are Watts / (m²- °K). We will use $\lambda = 1.9$ Watts / (m²- °K).

The heat capacity C of the Earth also has several components, the unit of C are (Watts – yr) / (m²- °K), we will use C = 40 (Watts – yr) / (m²- °K).

Q' is the additional atmospheric heating caused by increases in the level of CO_2, the units are Watts / m². Additional heat trapped in the atmosphere increases logarithmically with CO_2. Rewriting the equations, using Laplace transforms and taken partial fractions, we finally get:

$$T(t) = (0.0428\ \tau^2 / C)\ [\ t / \tau - (1 - e^{-t/\tau})\ ]$$

where $\tau = C / \lambda = 21.053$

Data: t, C, τ

Boundaries: t = 2010 To 2080 Step 10, or 10 To 80 Step 10, or 71 To 141 Step 10.

Total rows = n! . a . b . c . d . e = 120 (8) (1) (8) (1) (8) = 2,880

System of polynomials: the leading variable is years and years-increment, C and τ are constants. With x = 0.5 we select only eight polynomials as a group, showing that in 2080 the global warming will be 1.34 degree Celsius higher than today.

$$320\,x^5 + 640\,x^4 -\ 568\,x^3 + 84\,x^2 + 0.20\,x\ =\ 0.10$$
$$640\,x^5 + 640\,x^4 -\ 648\,x^3 + 84\,x^2 + 0.32\,x\ =\ 0.16$$
$$960\,x^5 + 640\,x^4 -\ 728\,x^3 + 84\,x^2 + 0.52\,x\ =\ 0.26$$
$$1280\,x^5 + 640\,x^4 -\ 808\,x^3 + 84\,x^2 + 0.96\,x\ =\ 0.48$$
$$1600\,x^5 + 640\,x^4 -\ 888\,x^3 + 84\,x^2 + 1.26\,x\ =\ 0.63$$
$$1920\,x^5 + 640\,x^4 -\ 968\,x^3 + 84\,x^2 + 1.74\,x\ =\ 0.87$$
$$2240\,x^5 + 640\,x^4 - 1048\,x^3 + 84\,x^2 + 2.14\,x\ =\ 1.07$$
$$\mathbf{2560\,x^5 + 640\,x^4 - 1128\,x^3 + 84\,x^2 + 2.68\,x\ =\ 1.34}$$

Case 2: Given very well known data, find the system of polynomials

THE TRAJECTORY OF AN ELEMENTARY PARTICLE

<u>Data</u>: Nuclear interaction produced by a cosmic-ray particle is photographed in a cloud chamber placed in a magnetic field. A heavy track is identified as a proton, and the coordinates of the track are the spatial coordinates X, Y, and Z. Some particle interactions and the magnetic field are known to determine this trajectory but not all, we will add a variable as additional cause for the total effect as unknown force A.

> Vector X: 1, 2, 3, 4
> Vector Y: 1, 2, 3, 5
> Vector Z: 1, 2, 4, 5
> Measured time = 0.09 at coordinates (3, 3, 4)

<u>Boundaries</u>:   X From 1 To 5 Step 1
Y From 1 To 5 Step 1
Z From 1 To 5 Step 1
t From 0.01 To 0.1 Step 0.02
A From 1.5 To 1.9 Step 0.2

Total rows: n! . a . b. c . d . e = 120 (5) (5) (5) (5) (3) = 225,000

<u>Analytic expression</u>:   None

<u>System of polynomials</u>: For each position (x, y, z) of the particle at time t, we investigate the unknown force of being between the boundaries given by A, the check point is the kinetic energy of the particle estimated at 11.99. We make the full run with x = 1 and sort the sum column looking for 11.99 at t = 0.09 for (3, 3, 4).

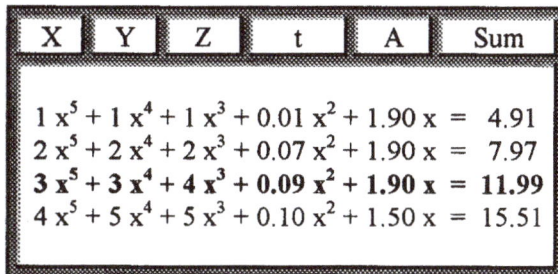

X	Y	Z	t	A	Sum
$1\,x^5 + 1\,x^4 + 1\,x^3 + 0.01\,x^2 + 1.90\,x$					4.91
$2\,x^5 + 2\,x^4 + 2\,x^3 + 0.07\,x^2 + 1.90\,x$					7.97
$\mathbf{3\,x^5 + 3\,x^4 + 4\,x^3 + 0.09\,x^2 + 1.90\,x}$					$\mathbf{11.99}$
$4\,x^5 + 5\,x^4 + 5\,x^3 + 0.10\,x^2 + 1.50\,x$					15.51

The equation in bold shows the unknown force A = 1.90

Gallery A: Tech and Science Graphs

This Gallery shows examples of simple,

complex, or symmetric output, the Mobius

ring can be found in one of them.

In particular Figure 54 is an example on how

to study unknown functions from places difficult

to reach, like the interior of atoms knowing

external, macro measurements.

Equally Figure 55 indicate that may be possible

to find the developments inside a living cell in

medicine studies, or the cell itself:

Gallery B: Graph 19.

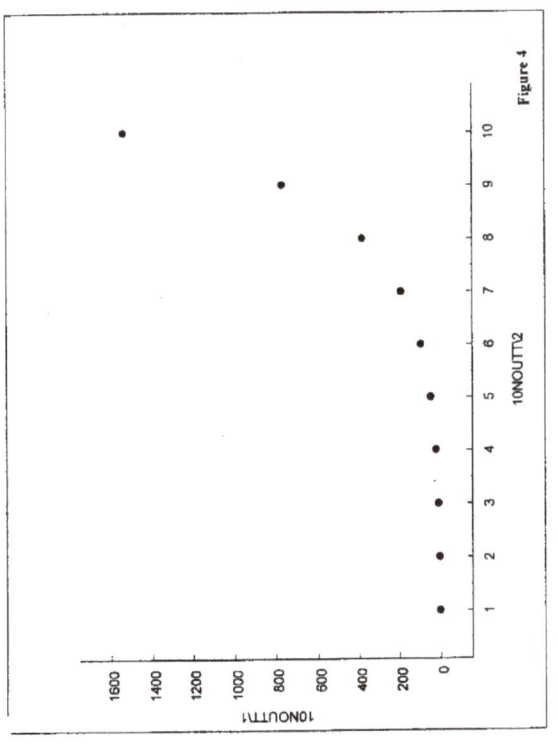

Figure 2

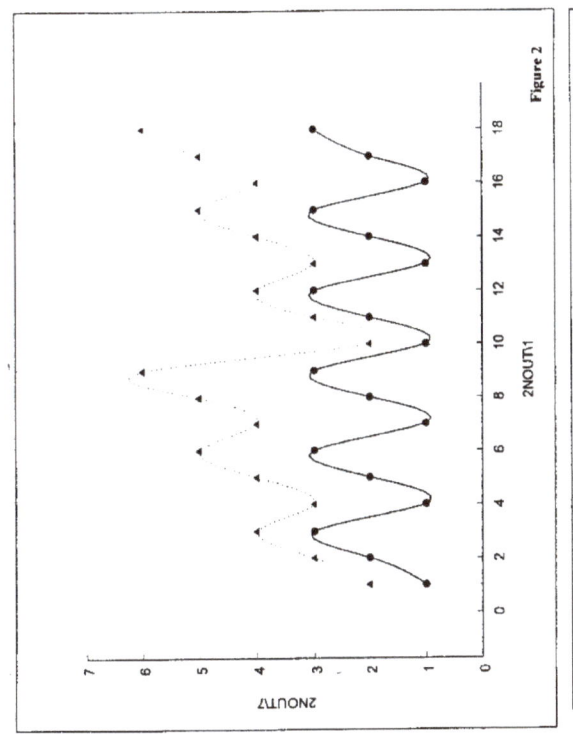

Figure 1

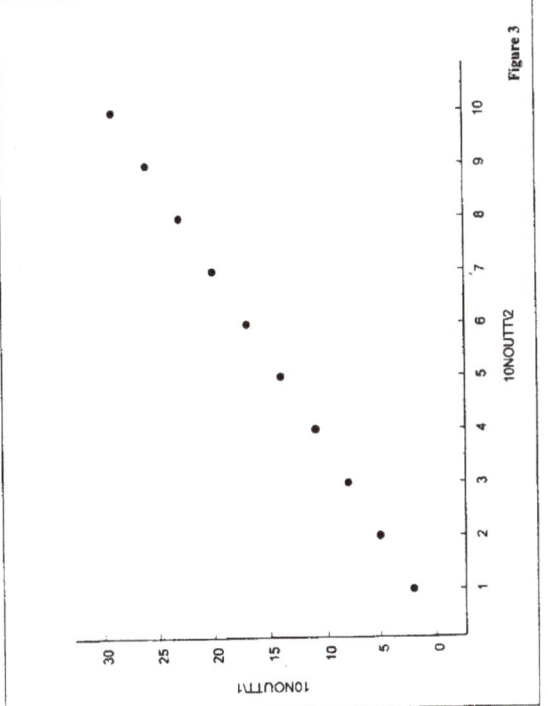

Figure 4

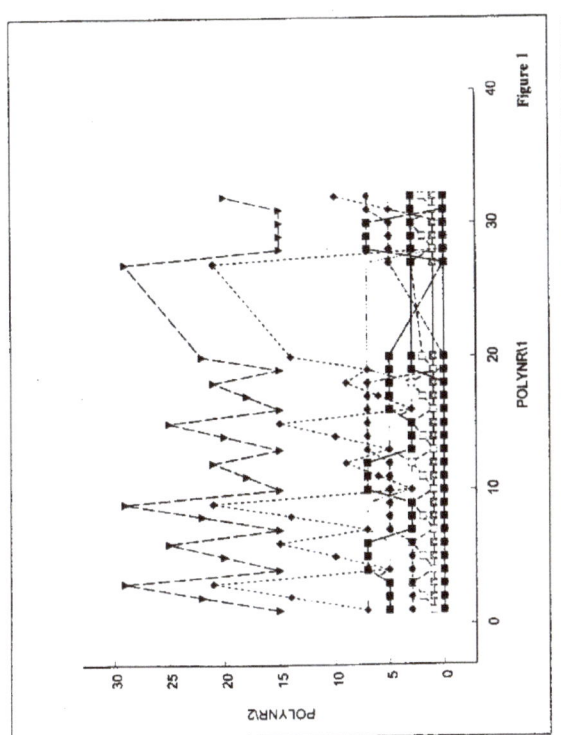

Figure 3

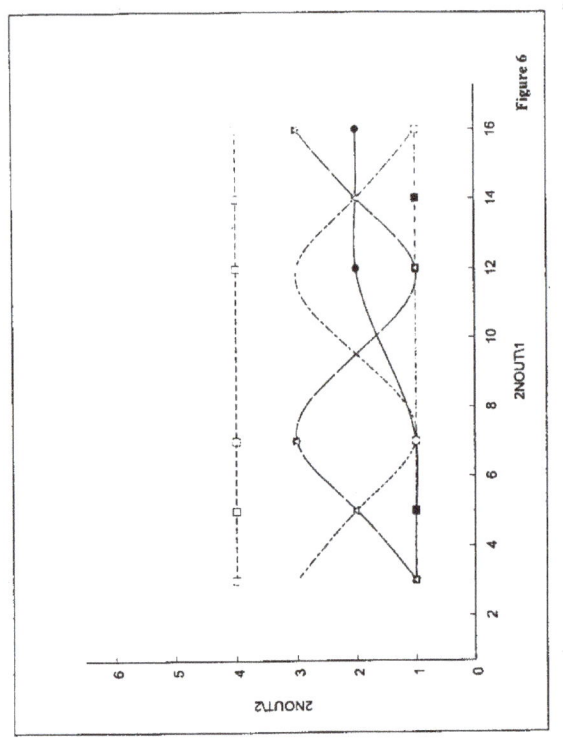

Figure 6

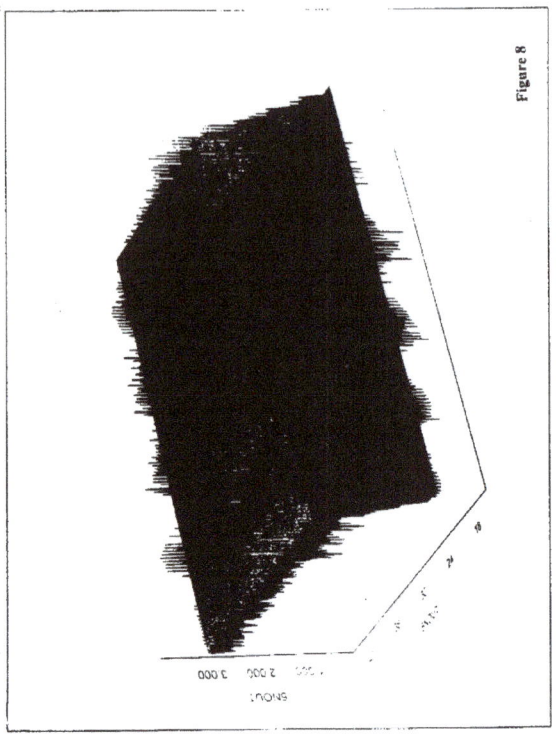

Figure 8

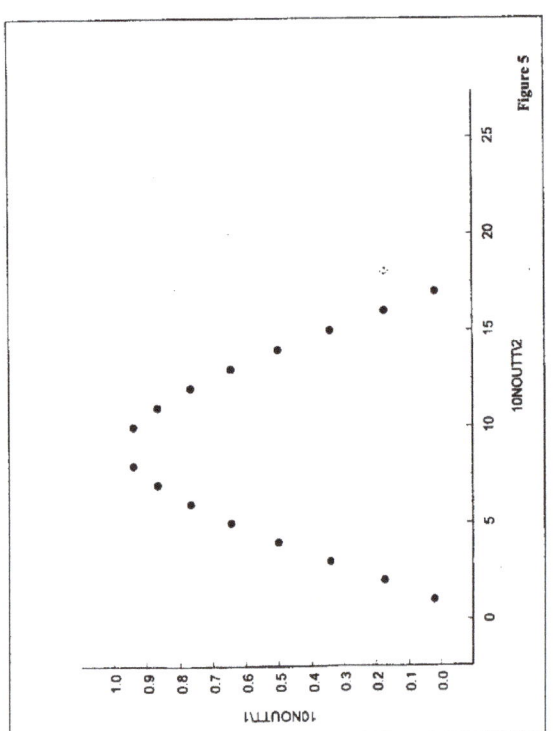

Figure 5

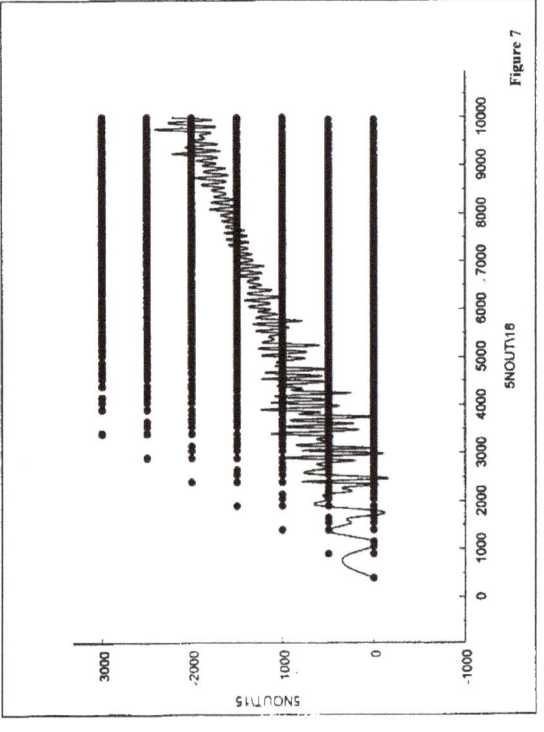

Figure 7

Figure 10

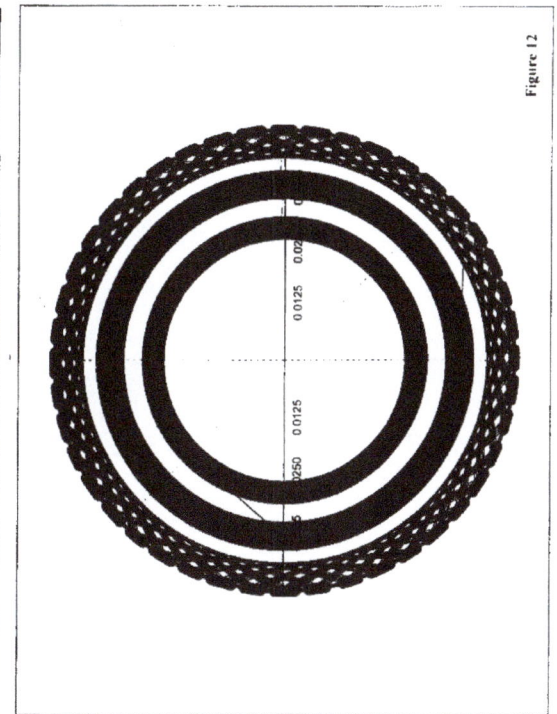

Figure 12

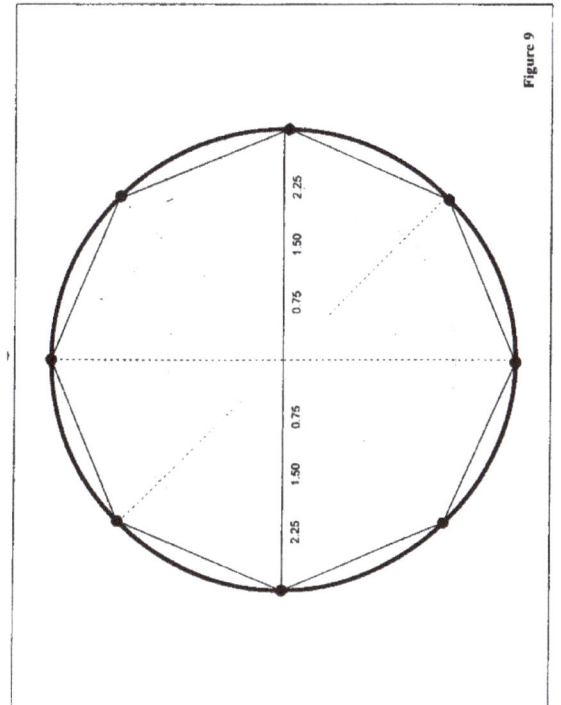

Figure 9

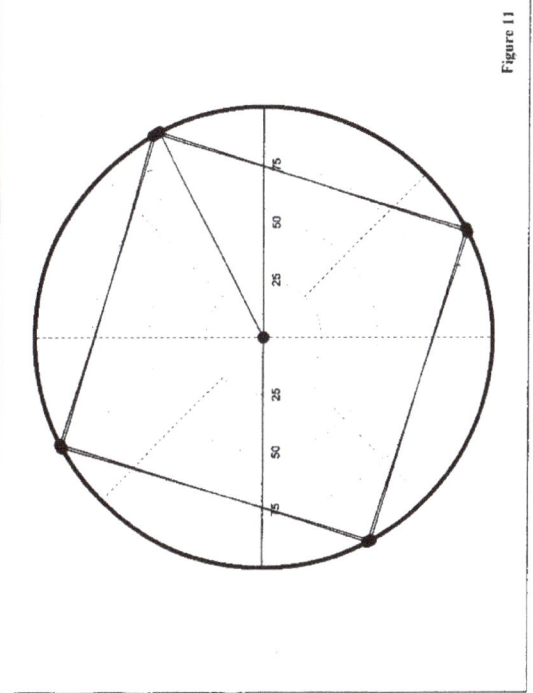

Figure 11

82

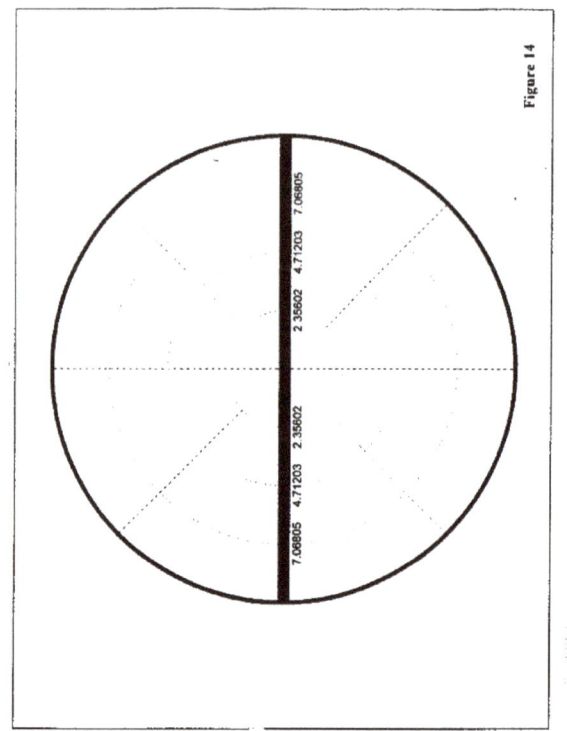

Figure 14

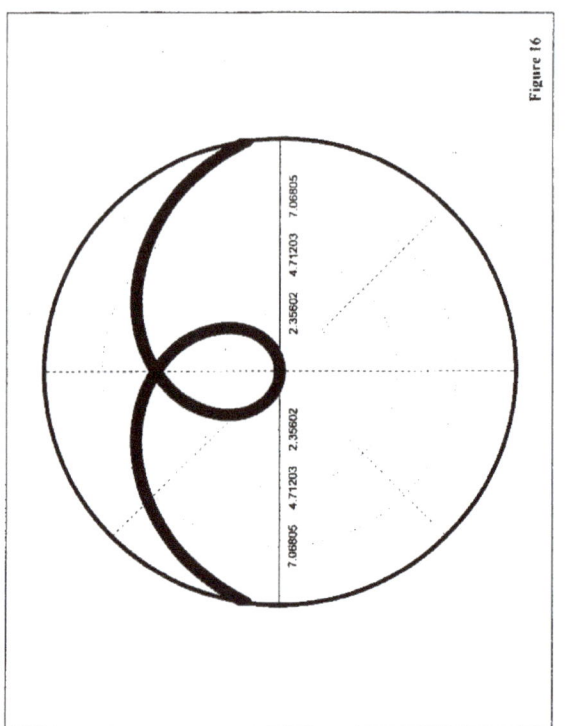

Figure 16

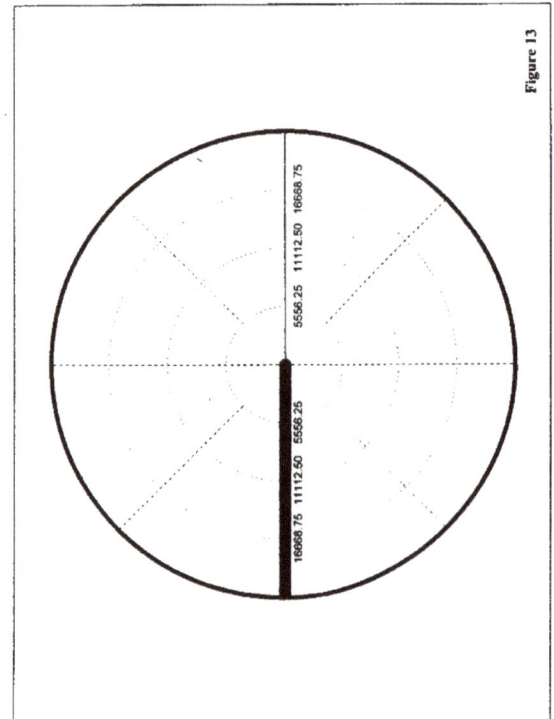

Figure 13

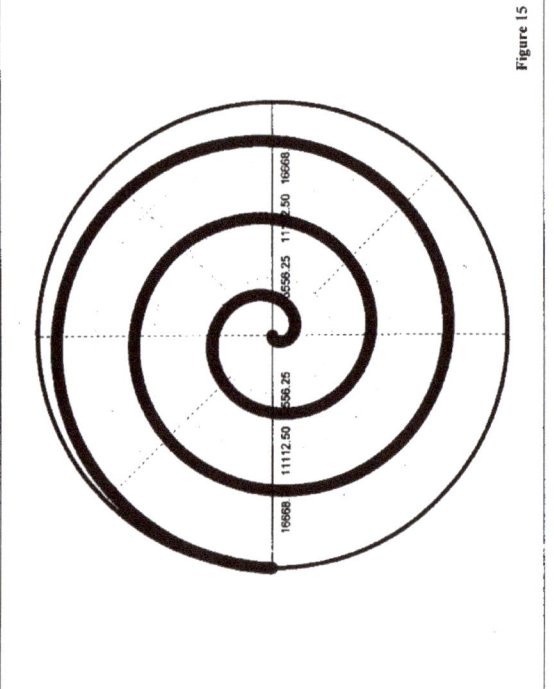

Figure 15

83

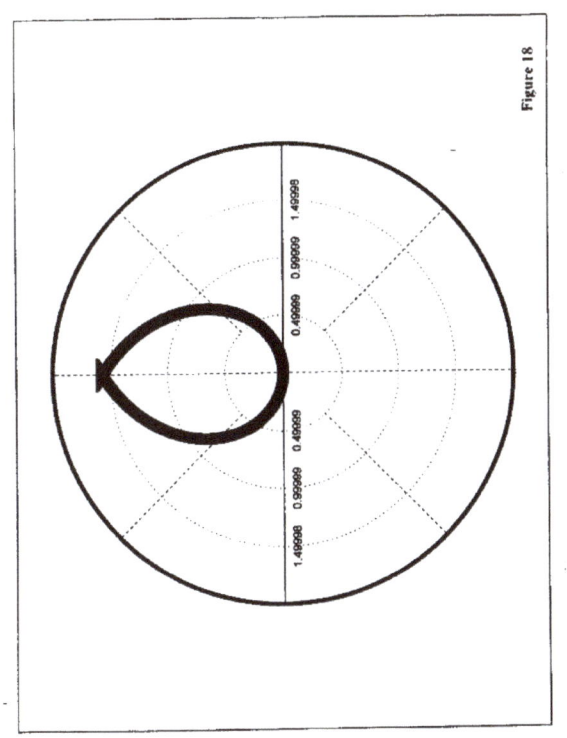

Figure 18

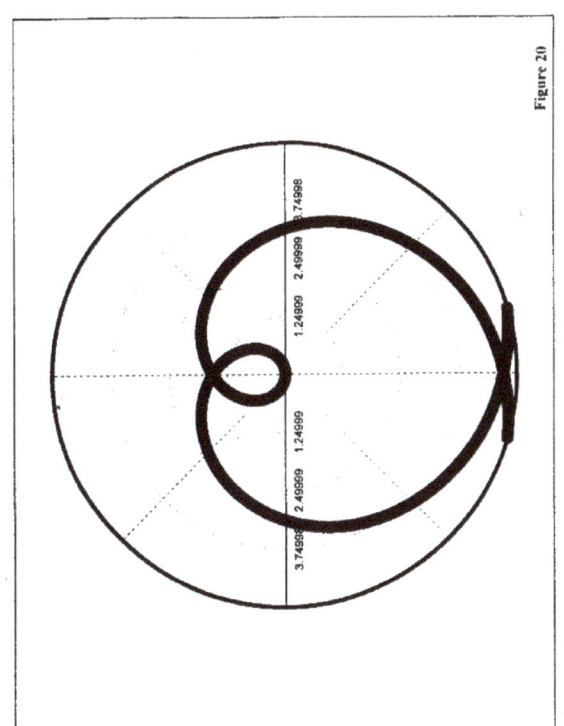

Figure 20

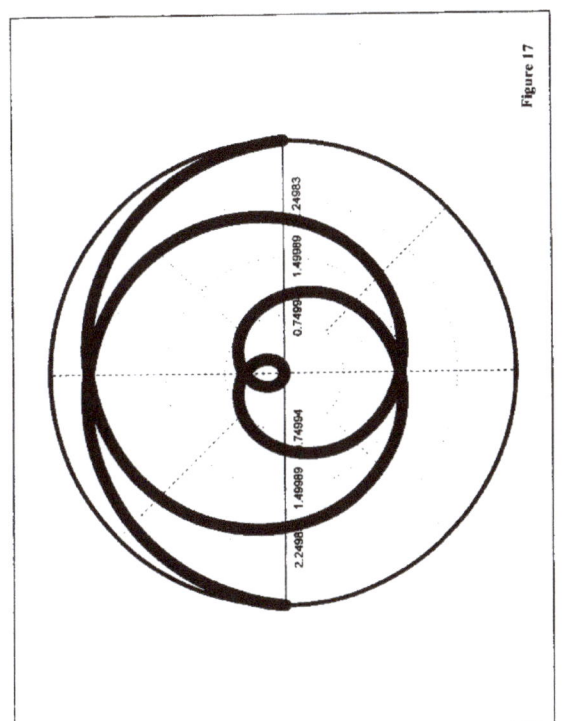

Figure 17

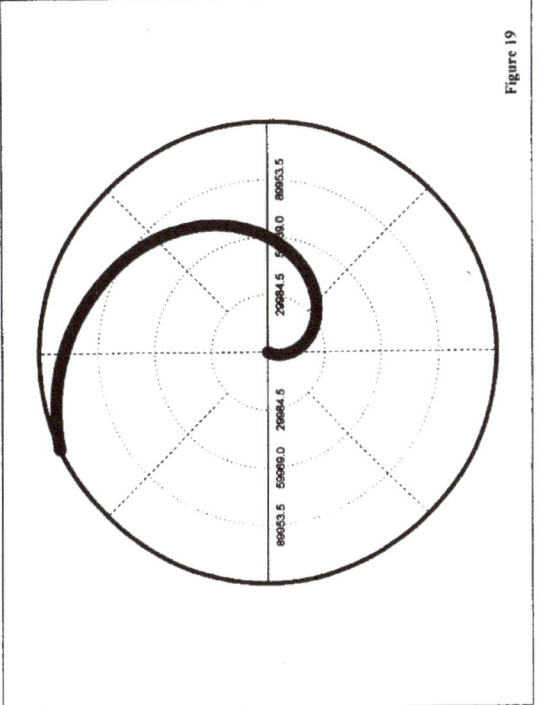

Figure 19

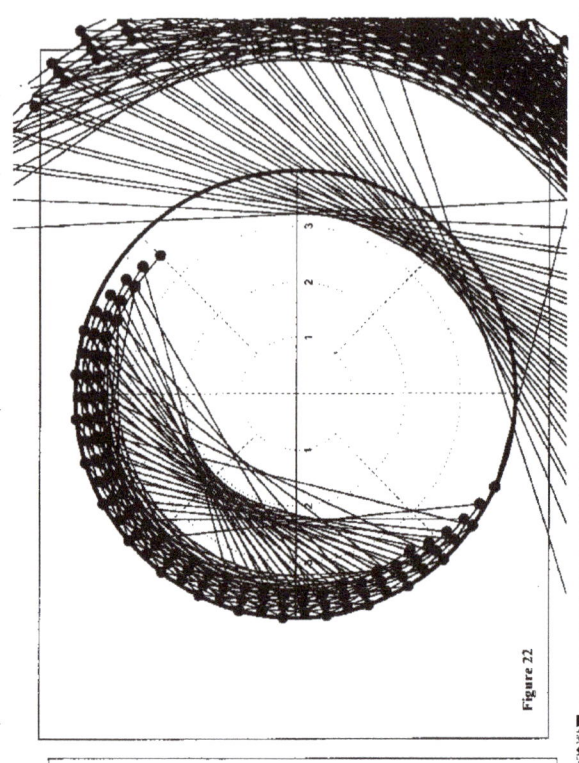

Figure 22

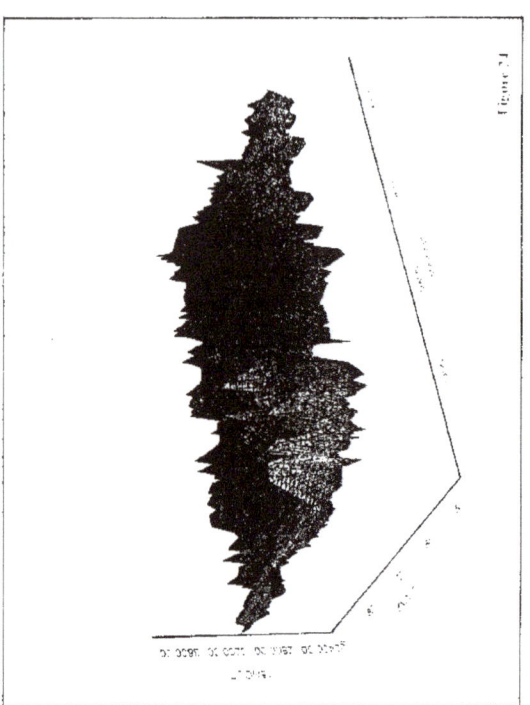

Figure 21

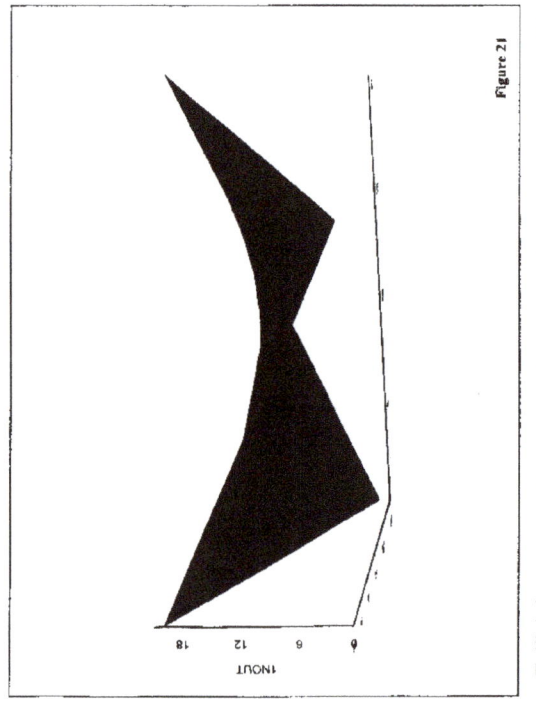

Figure 21

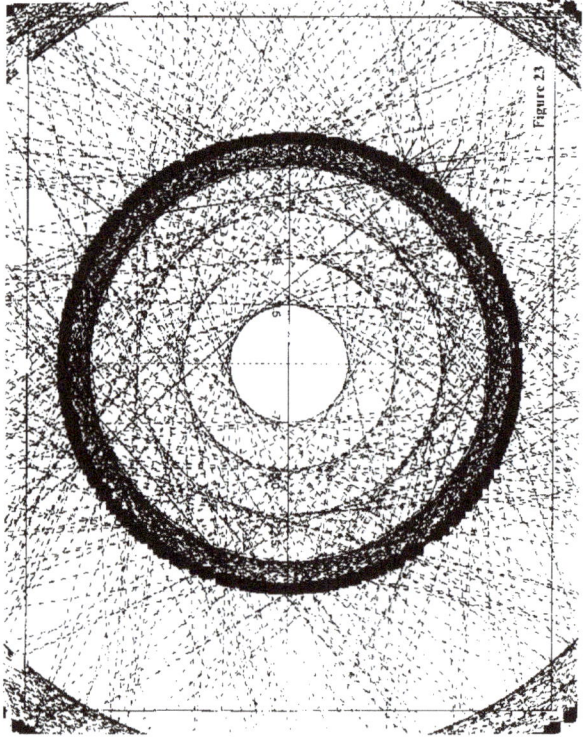

Figure 23

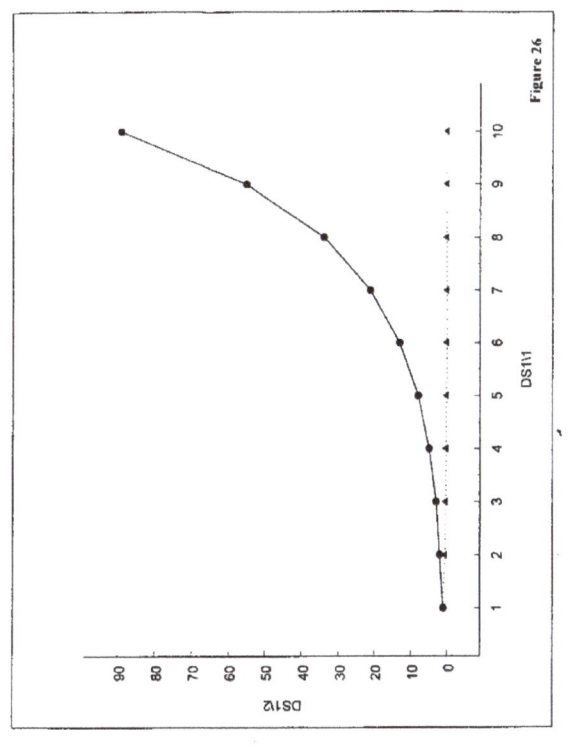

Figure 26

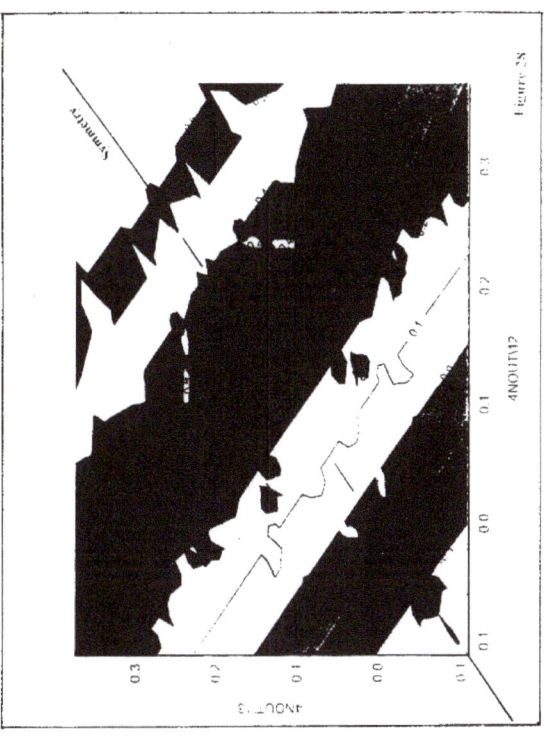

Figure 28

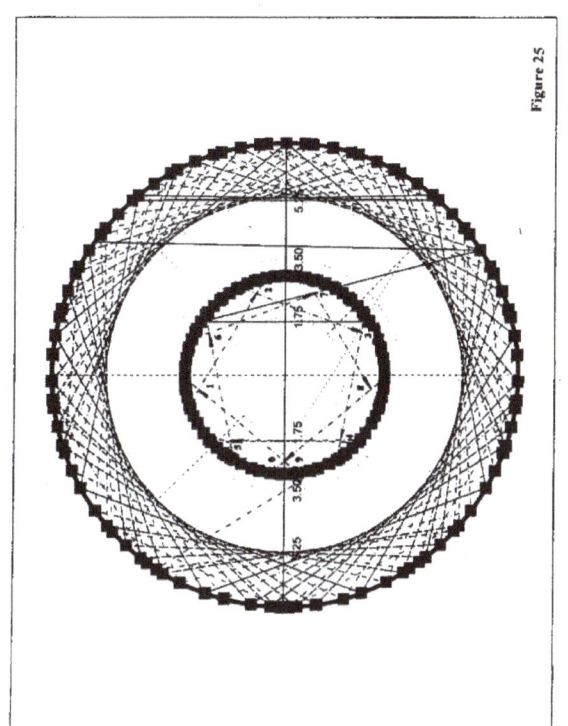

Figure 25

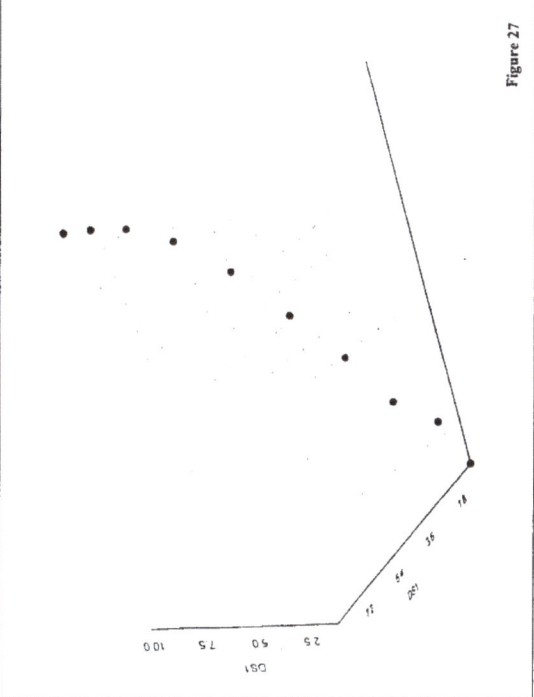

Figure 27

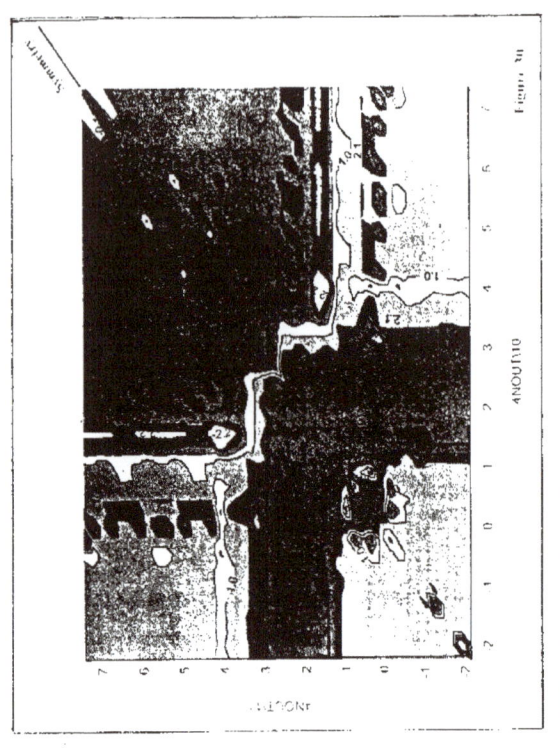

Figure 30

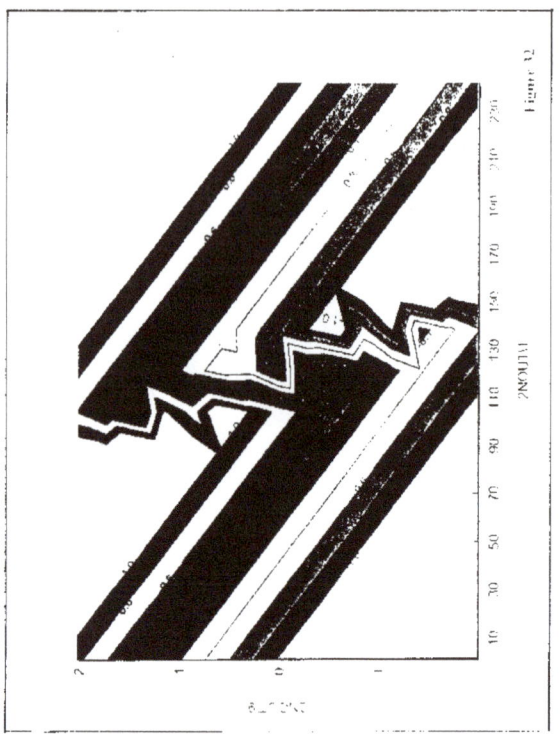

Figure 32

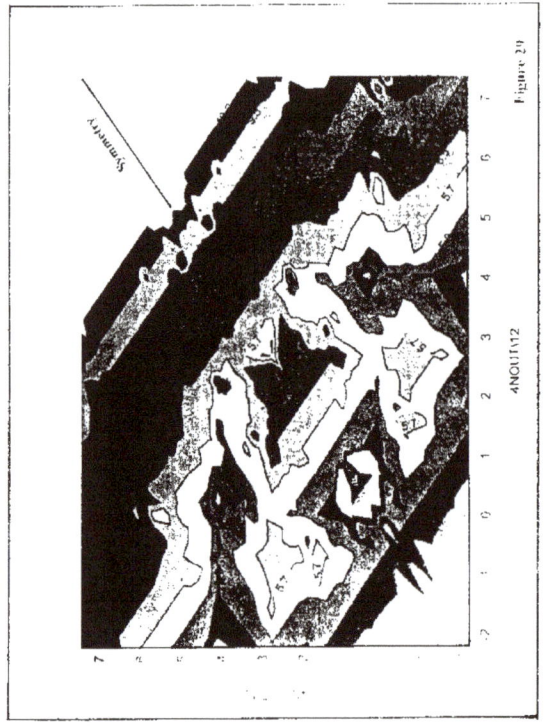

Figure 29

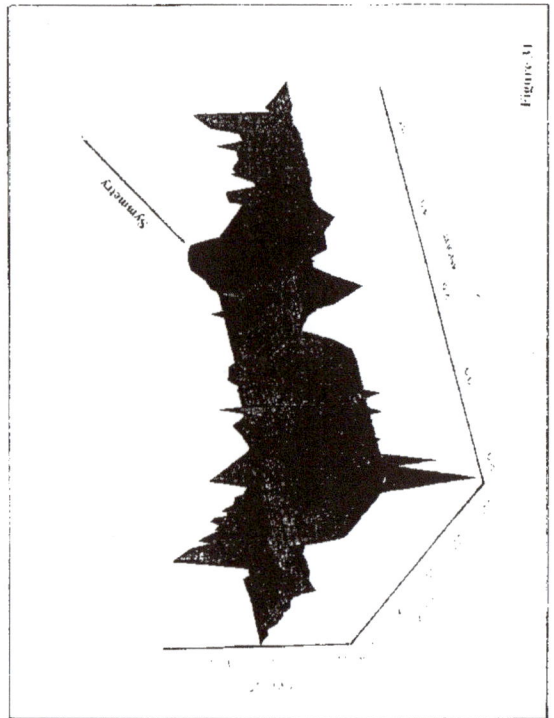

Figure 31

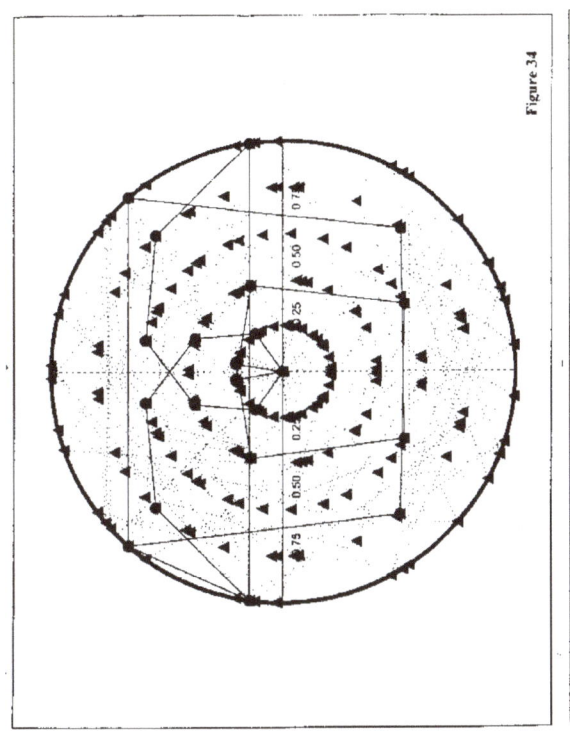

Figure 34

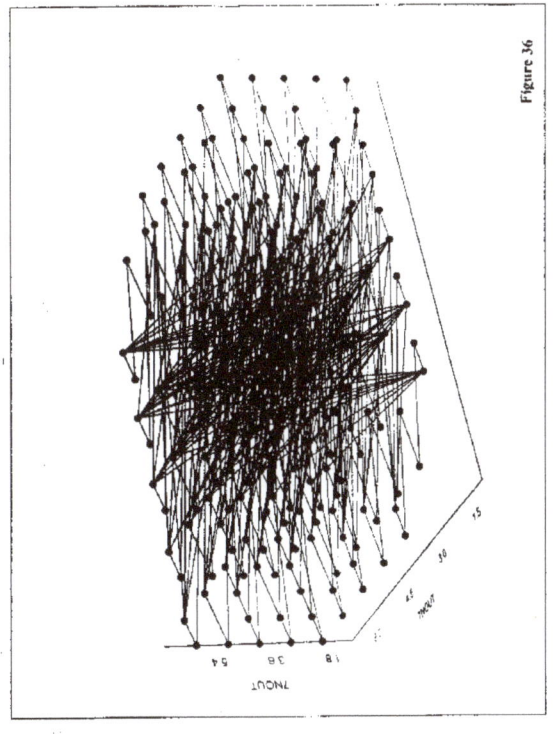

Figure 36

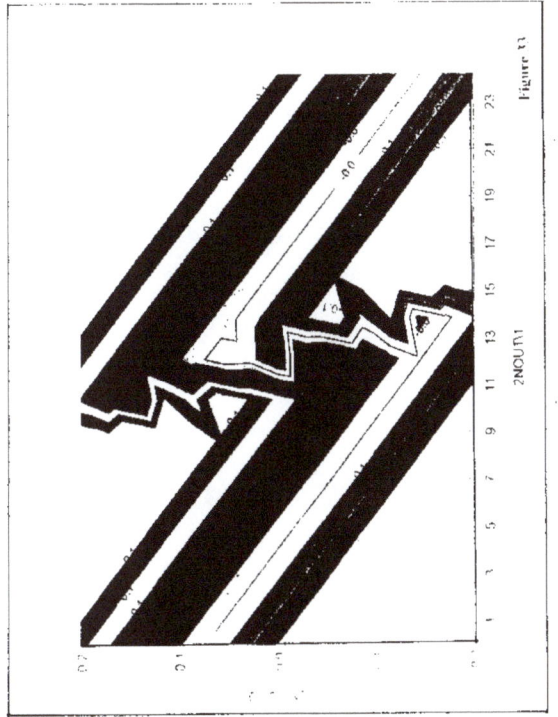

Figure 33

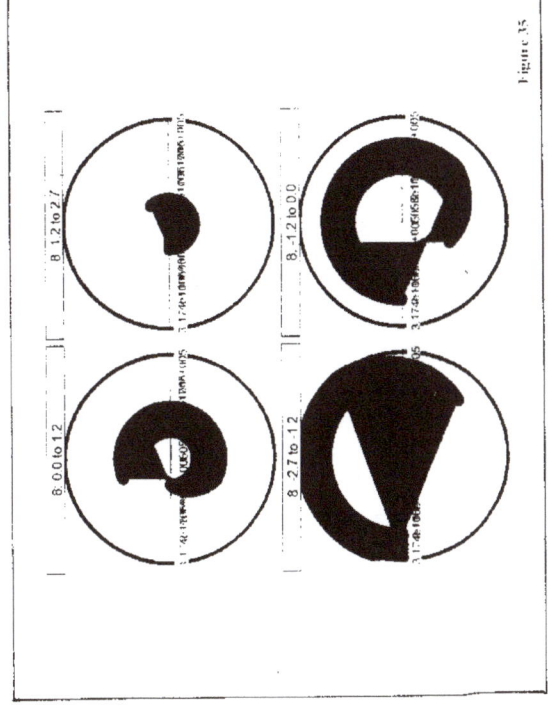

Figure 35

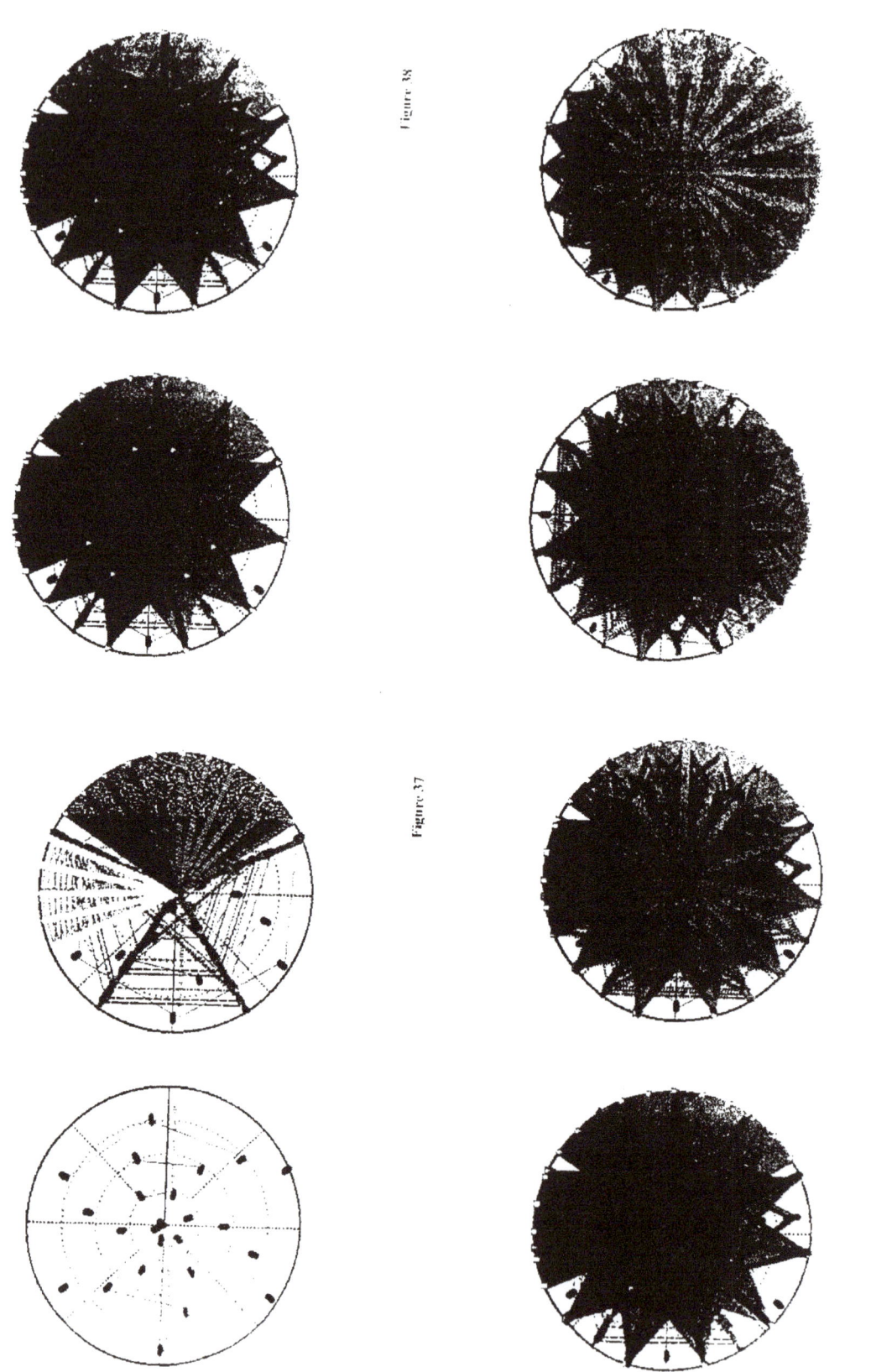

Figure 37

Figure 38

Figure 39

Figure 40

Figure 42

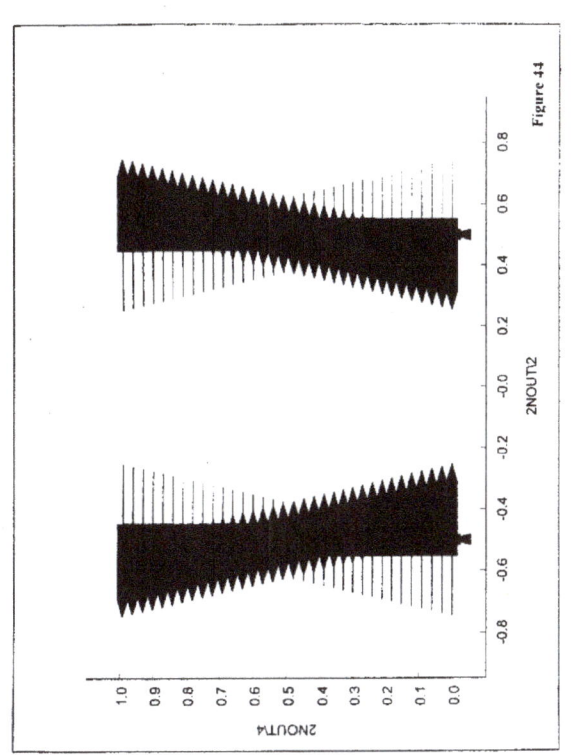

Figure 44

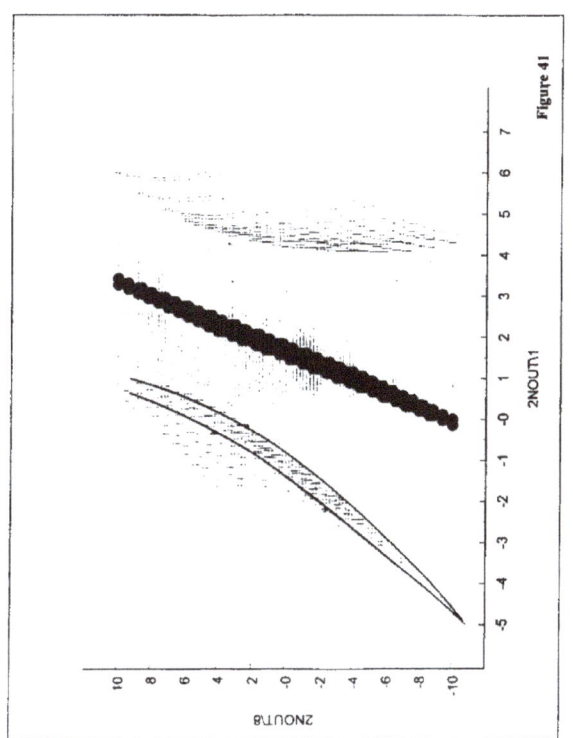

Figure 41

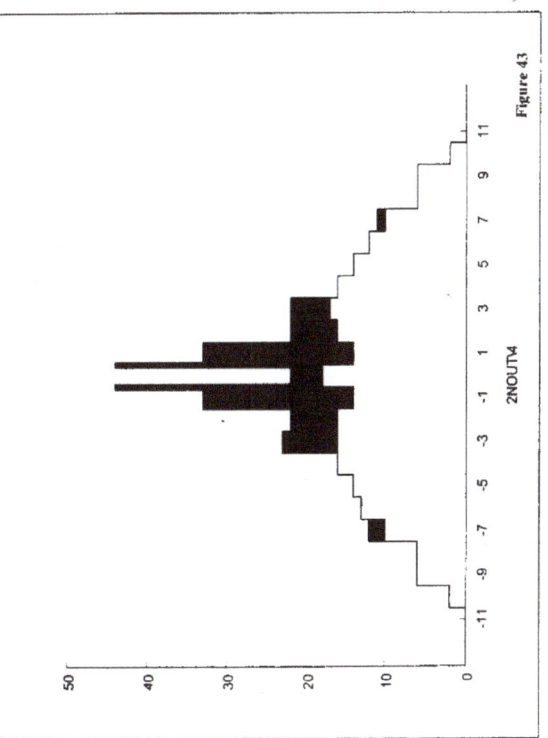

Figure 43

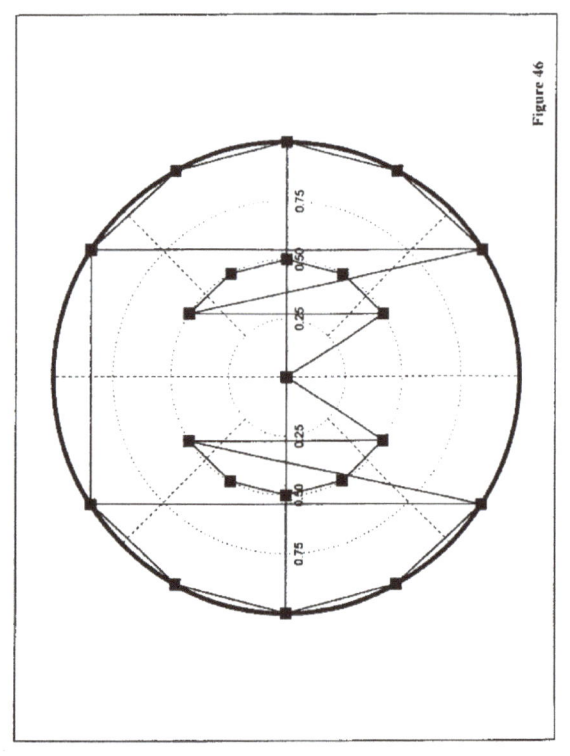

Figure 46

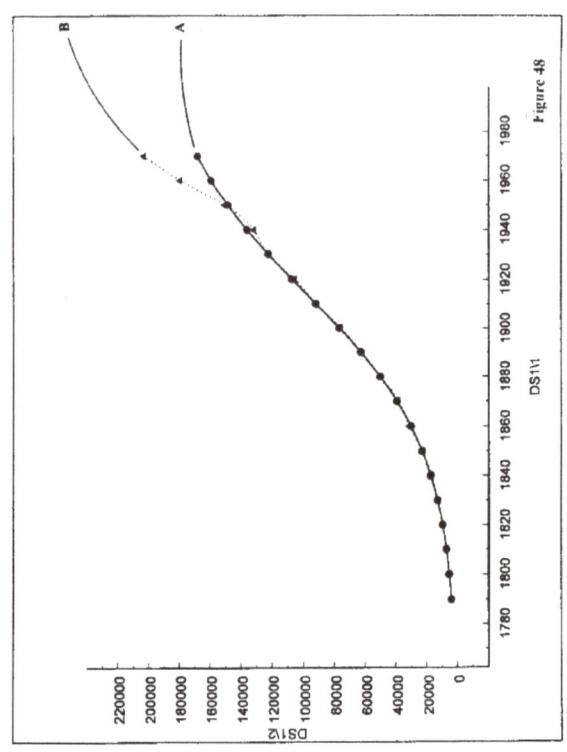

Figure 48

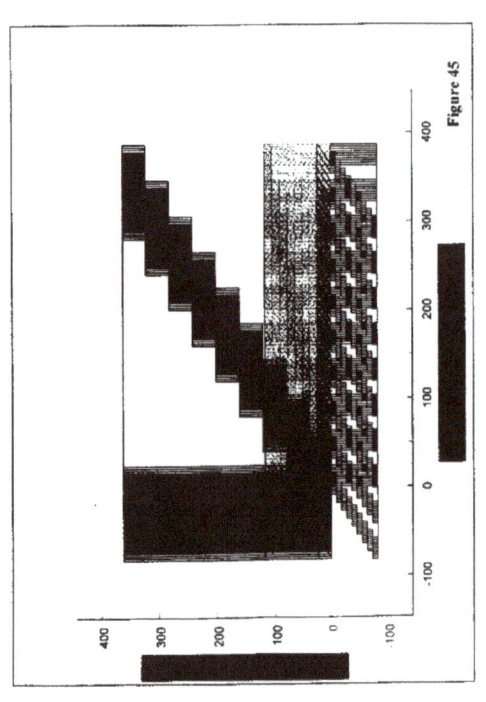

Figure 45

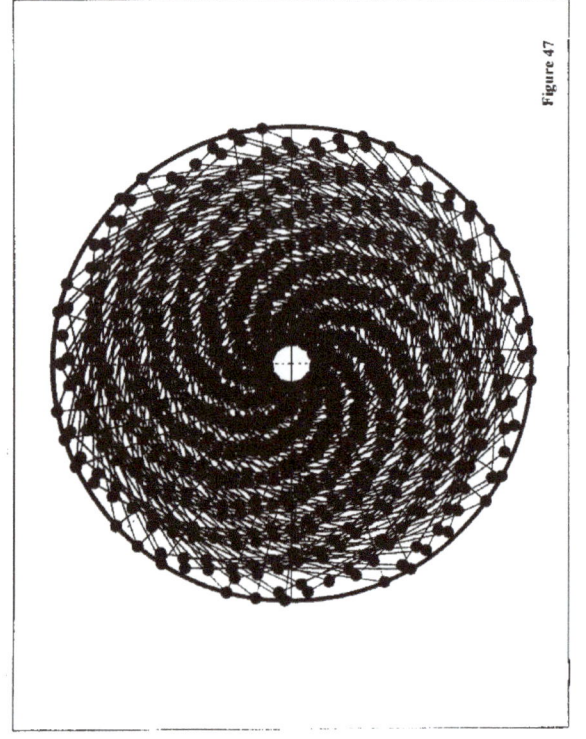

Figure 47

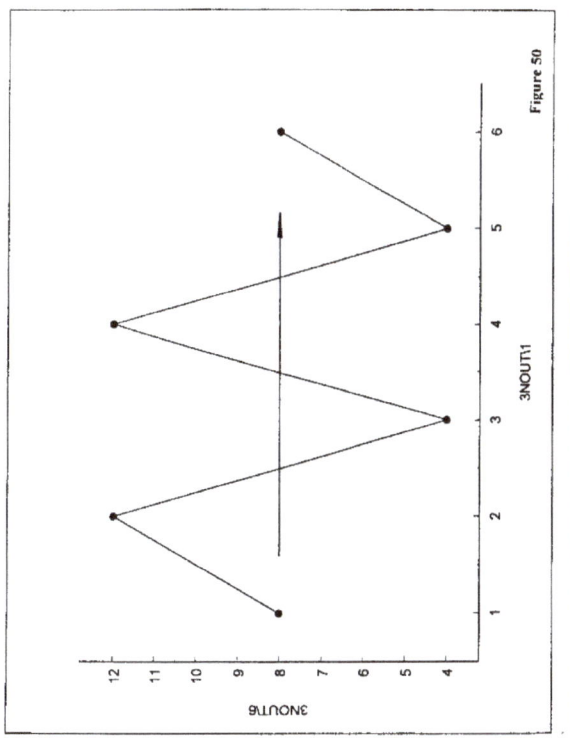

Figure 49

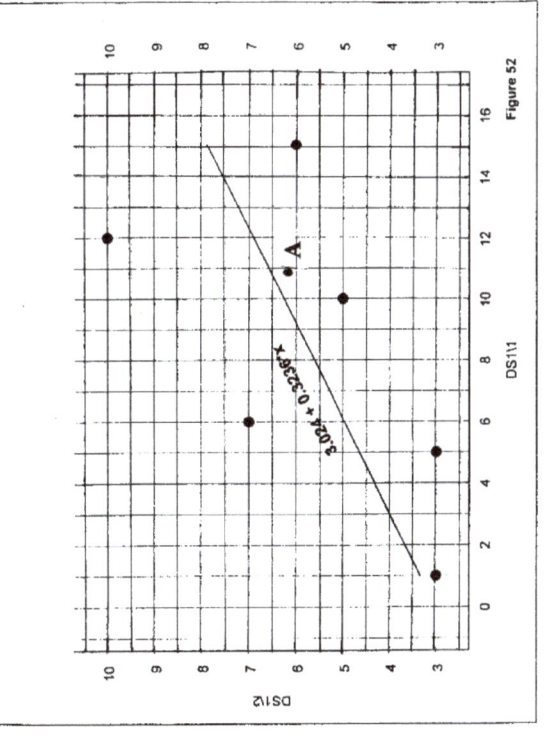

Figure 50

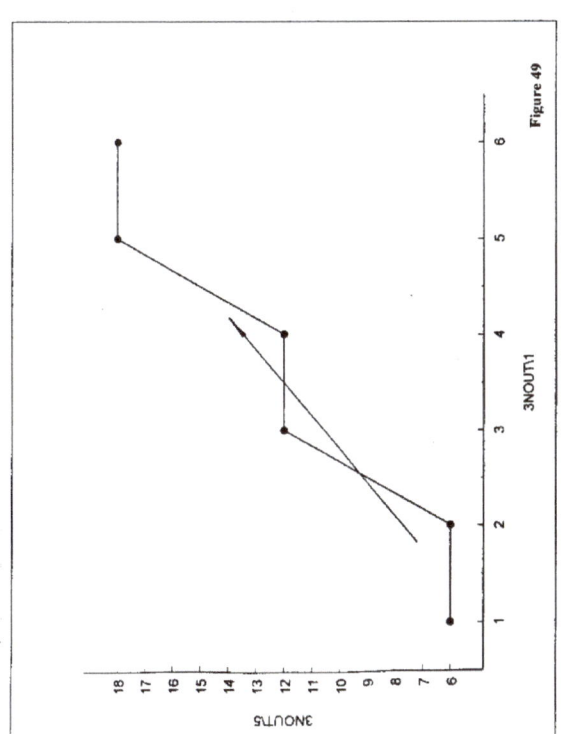

Figure 51

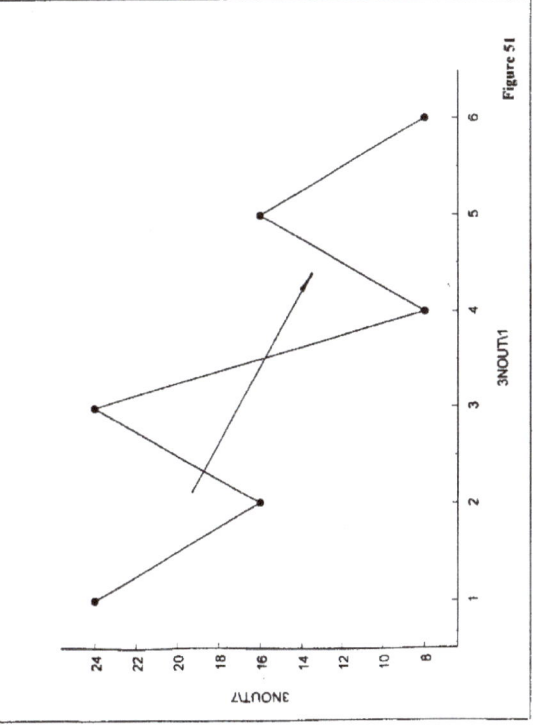

Figure 52

The beginning of portraiture....

(Use MathCad grays scale)

$$80\,x^{7} + 80\,x^{6} + 80\,x^{5} + 80\,x^{4} + 80\,x^{3} + 80\,x^{2} + 80\,x^{1} = 560$$
$$80\,x^{7} + 220\,x^{6} + 240\,x^{5} + 100\,x^{4} + 240\,x^{3} + 220\,x^{2} + 80\,x^{1} = 1180$$
$$80\,x^{7} + 90\,x^{6} + 100\,x^{5} + 120\,x^{4} + 100\,x^{3} + 90\,x^{2} + 80\,x^{1} = 660$$
$$80\,x^{7} + 90\,x^{6} + 100\,x^{5} + 160\,x^{4} + 100\,x^{3} + 90\,x^{2} + 80\,x^{1} = 700$$
$$80\,x^{7} + 90\,x^{6} + 100\,x^{5} + 100\,x^{4} + 100\,x^{3} + 90\,x^{2} + 80\,x^{1} = 640$$
$$80\,x^{7} + 90\,x^{6} + 200\,x^{5} + 200\,x^{4} + 200\,x^{3} + 90\,x^{2} + 80\,x^{1} = 940$$
$$80\,x^{7} + 80\,x^{6} + 80\,x^{5} + 80\,x^{4} + 80\,x^{3} + 80\,x^{2} + 80\,x^{1} = 560$$
$$x = 1$$

Figure 53

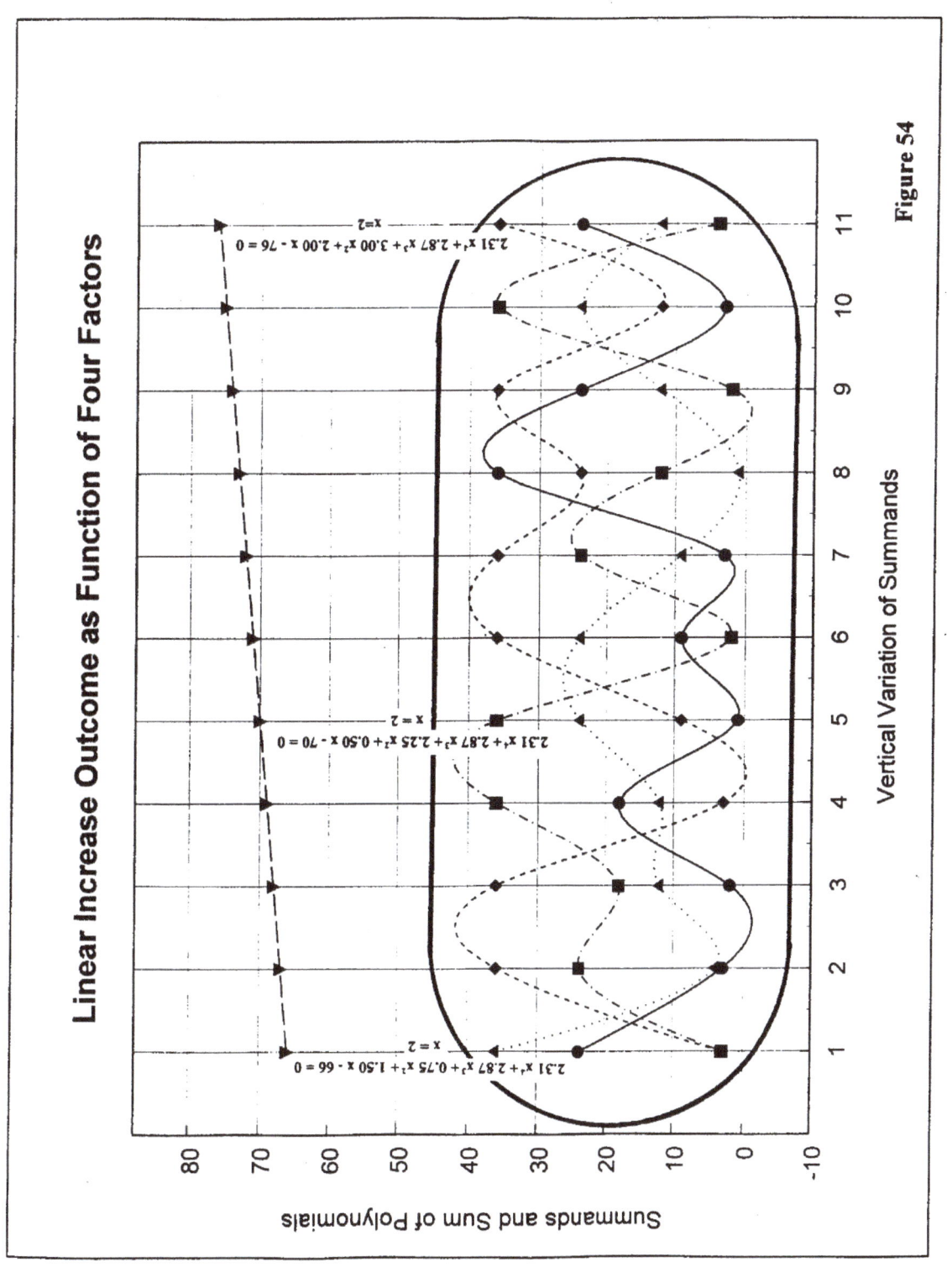

Linear Increase Outcome as Function of Four Factors

$2.31\ x^4 + 2.87\ x^3 + 3.00\ x^2 + 2.00\ x - 76 = 0$ x = 2

$2.31\ x^4 + 2.87\ x^3 + 2.25\ x^2 + 0.50\ x - 70 = 0$ x = 2

$2.31\ x^4 + 2.87\ x^3 + 0.75\ x^2 + 1.50\ x - 66 = 0$ x = 2

Summands and Sum of Polynomials

Vertical Variation of Summands

Figure 54

94

Example of standard outcome as function of four trend factors

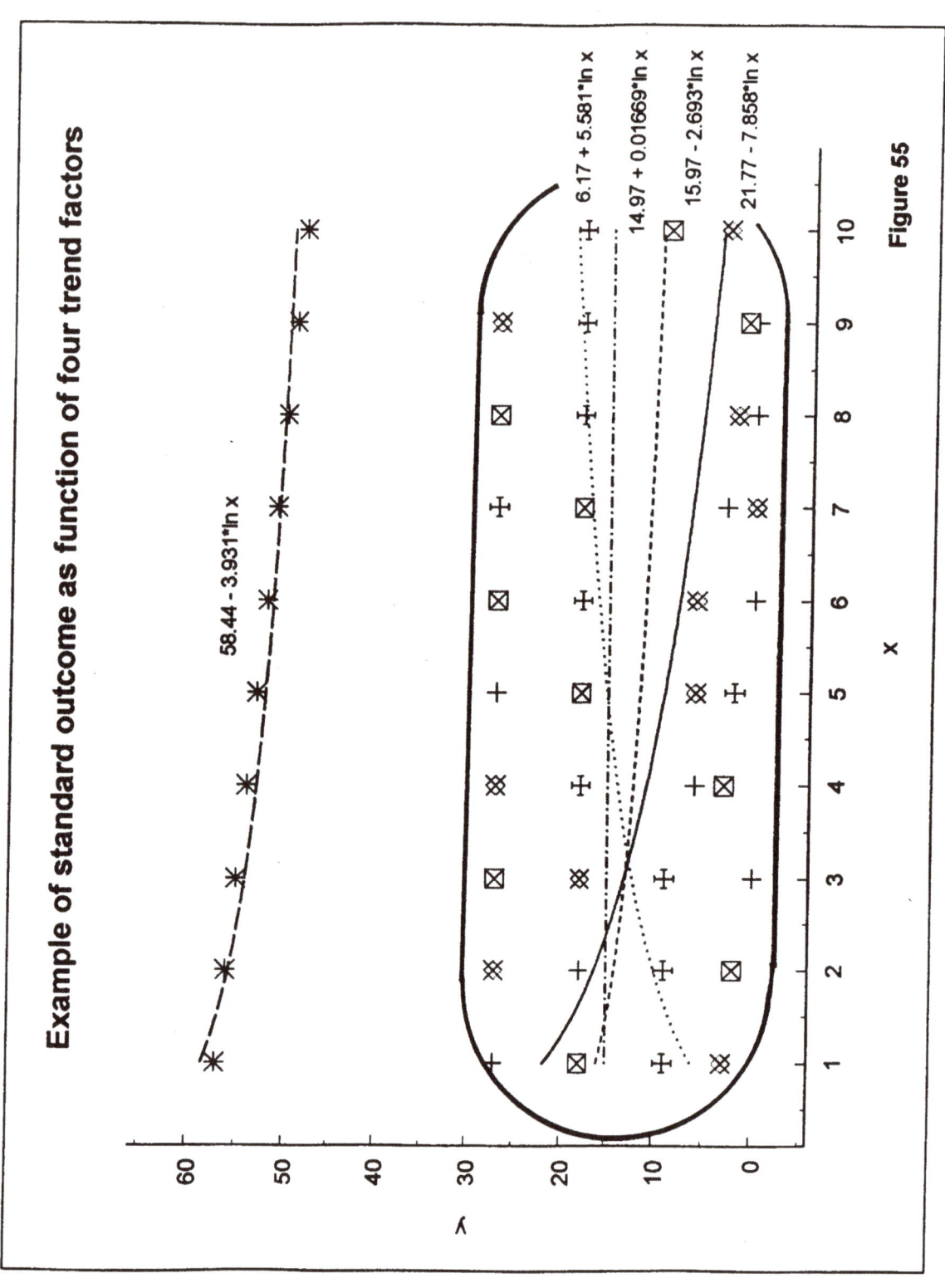

58.44 - 3.931*ln x

6.17 + 5.581*ln x

14.97 + 0.01669*ln x

15.97 - 2.693*ln x

21.77 - 7.858*ln x

Figure 55

Gallery B: Representative Art

Here we find, in a fly of the imagination, the sky with a round body and stars, structural trusses, in others simple beauty.

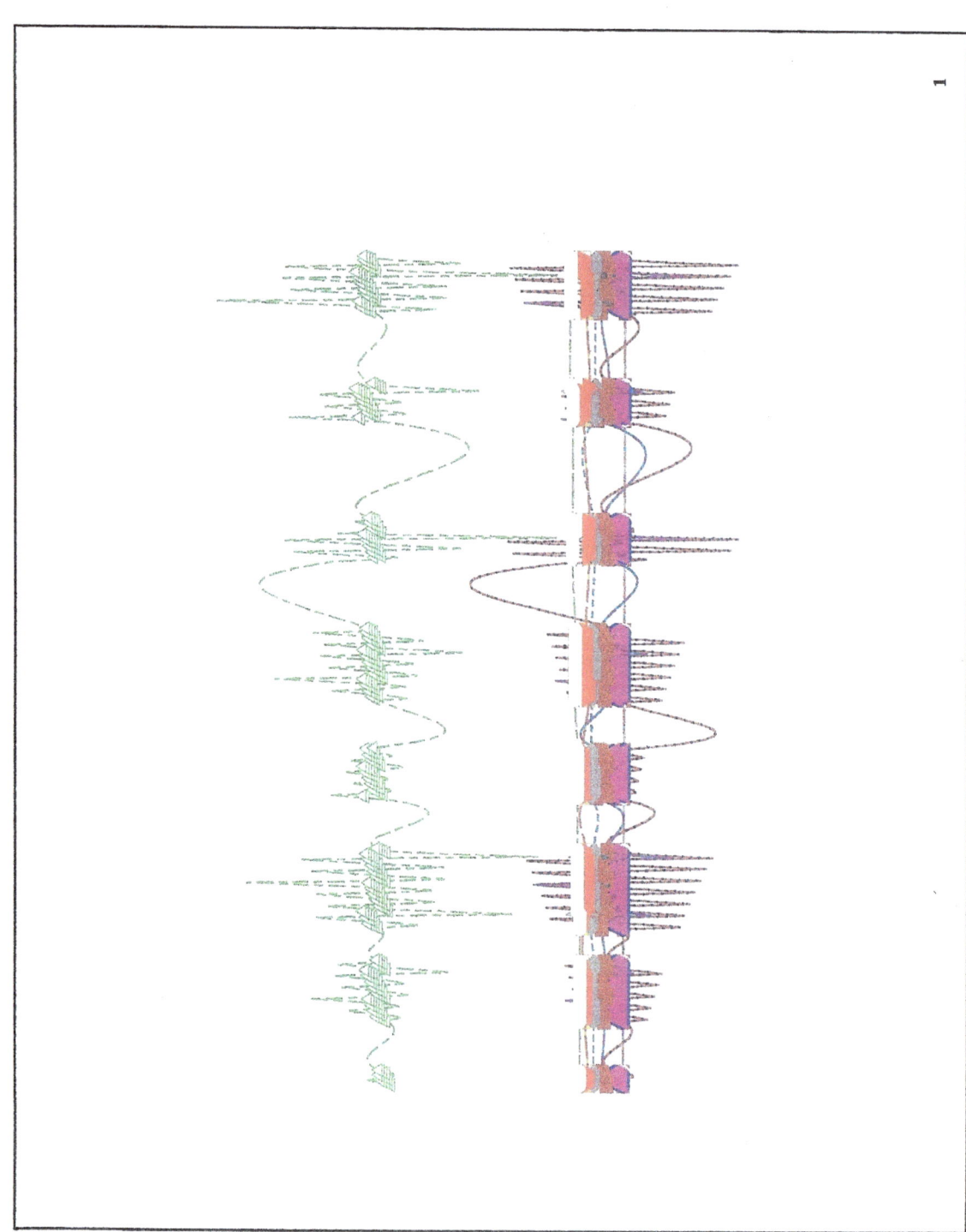

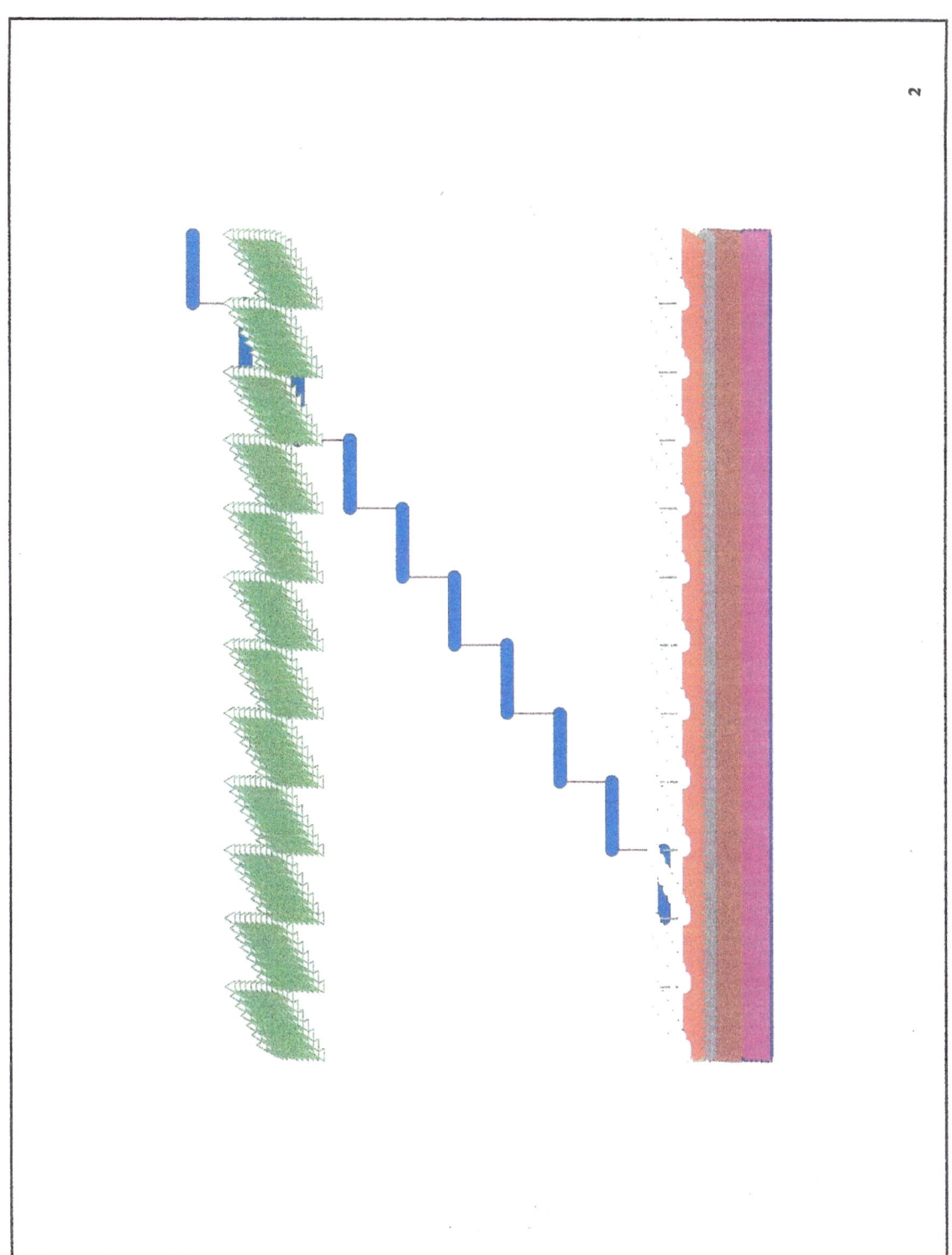

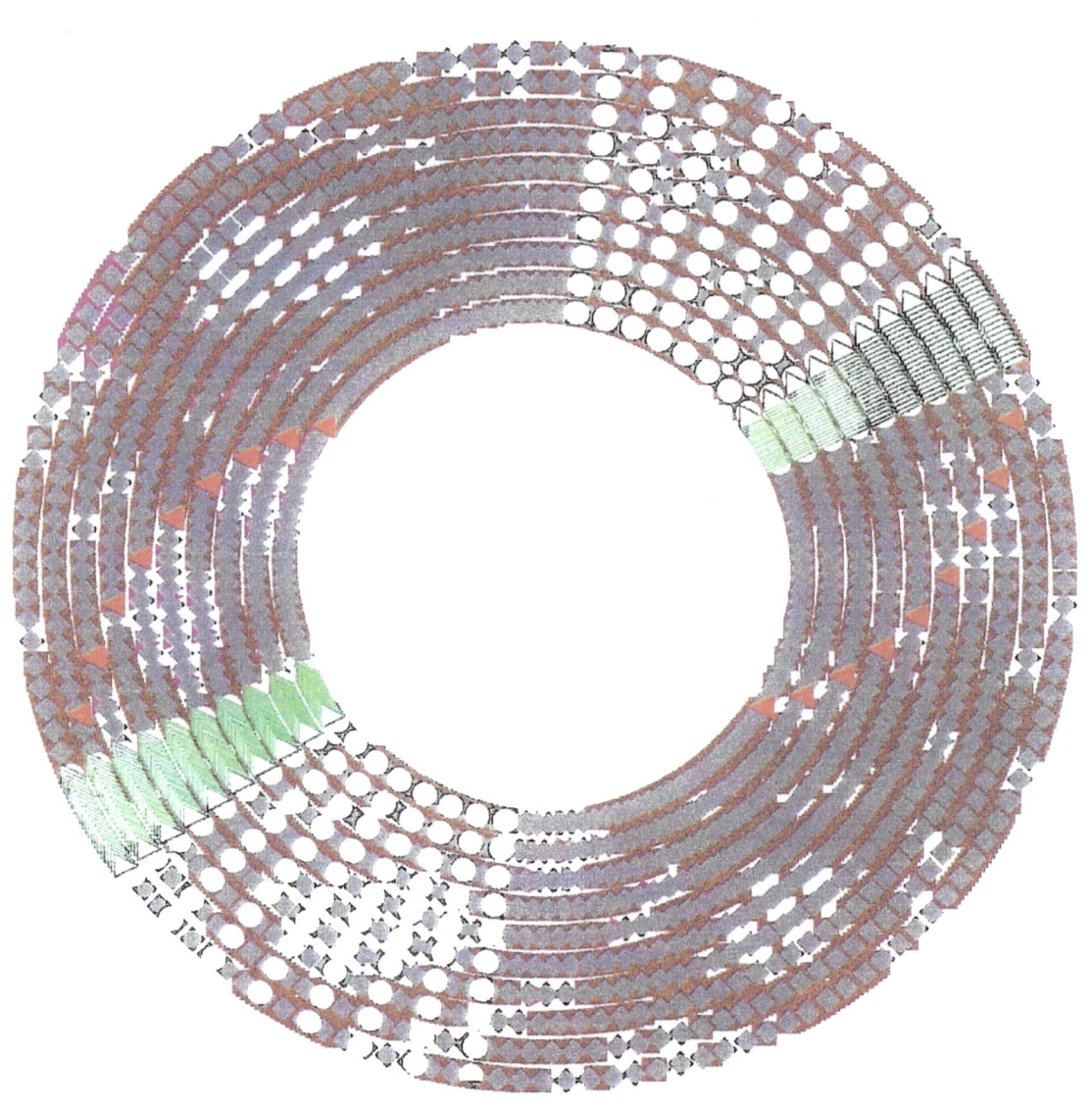

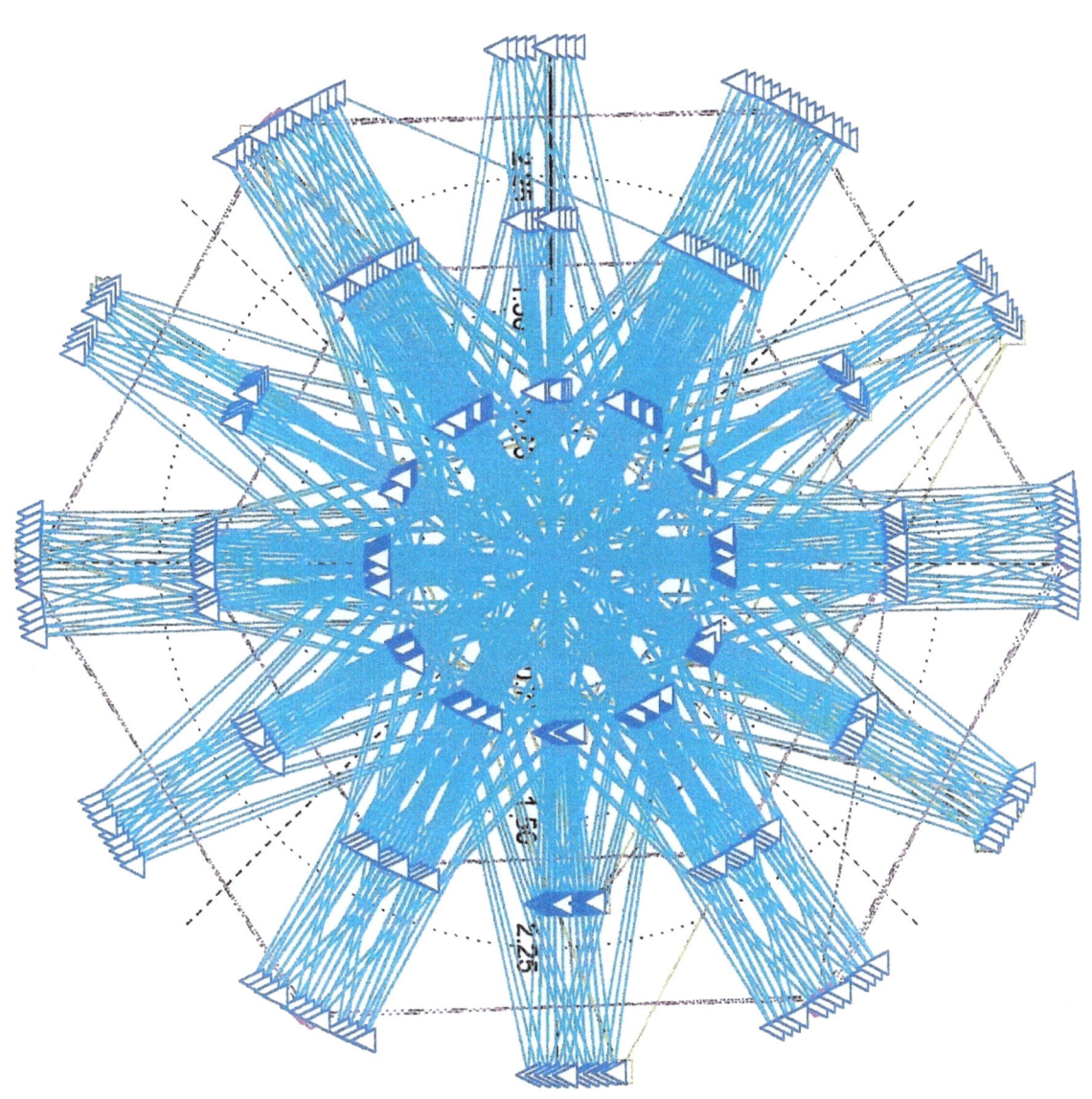

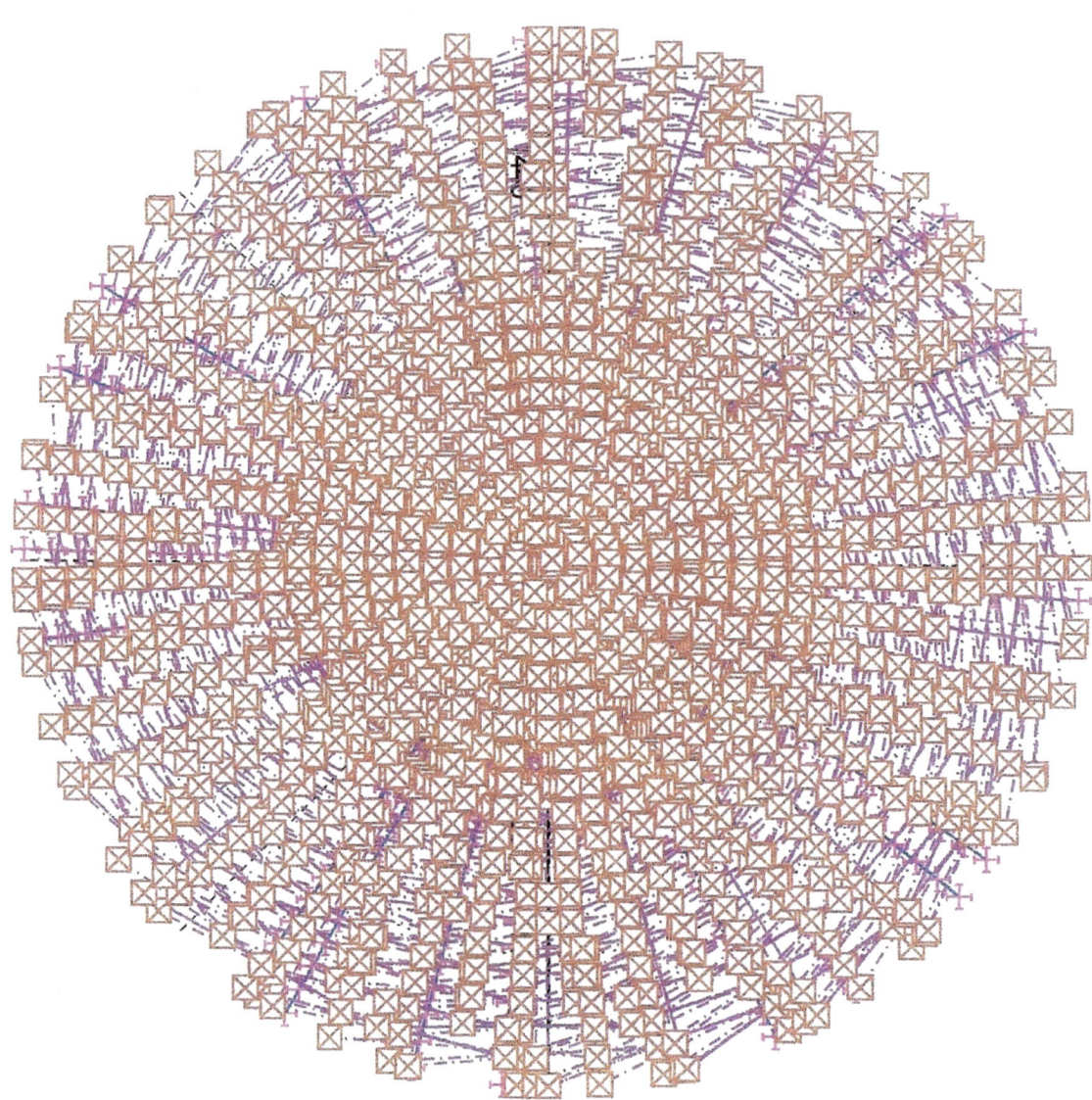

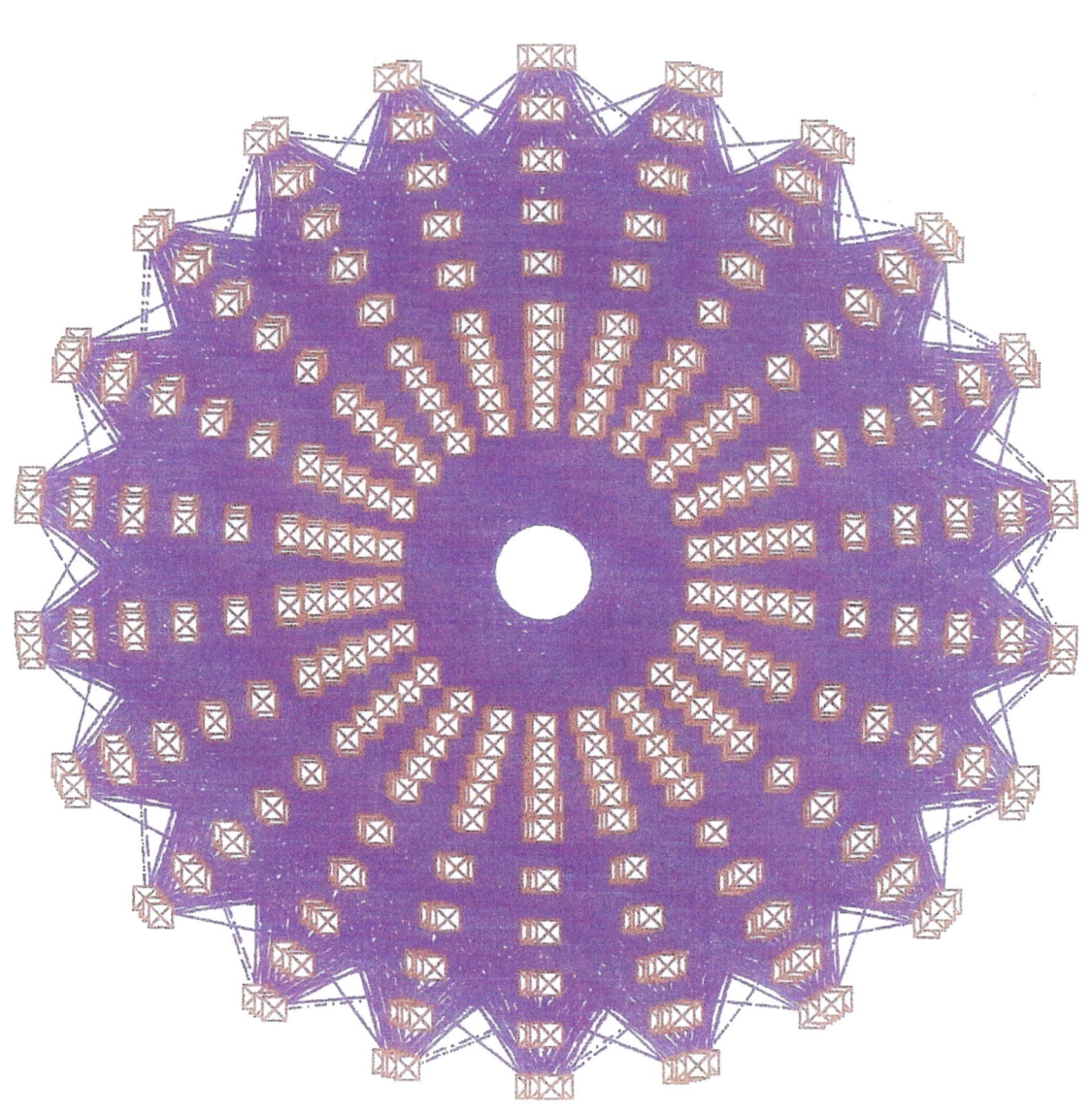

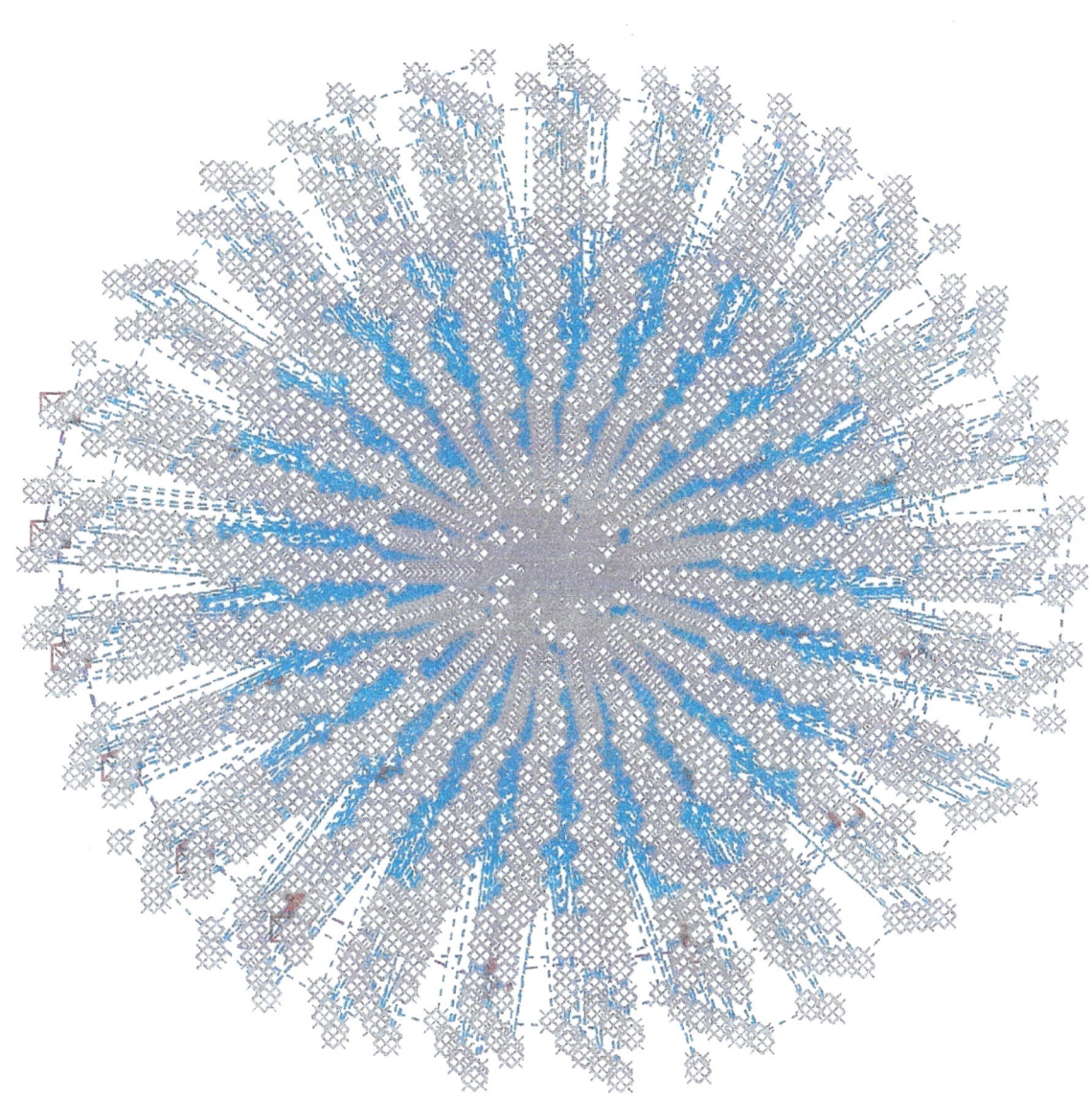

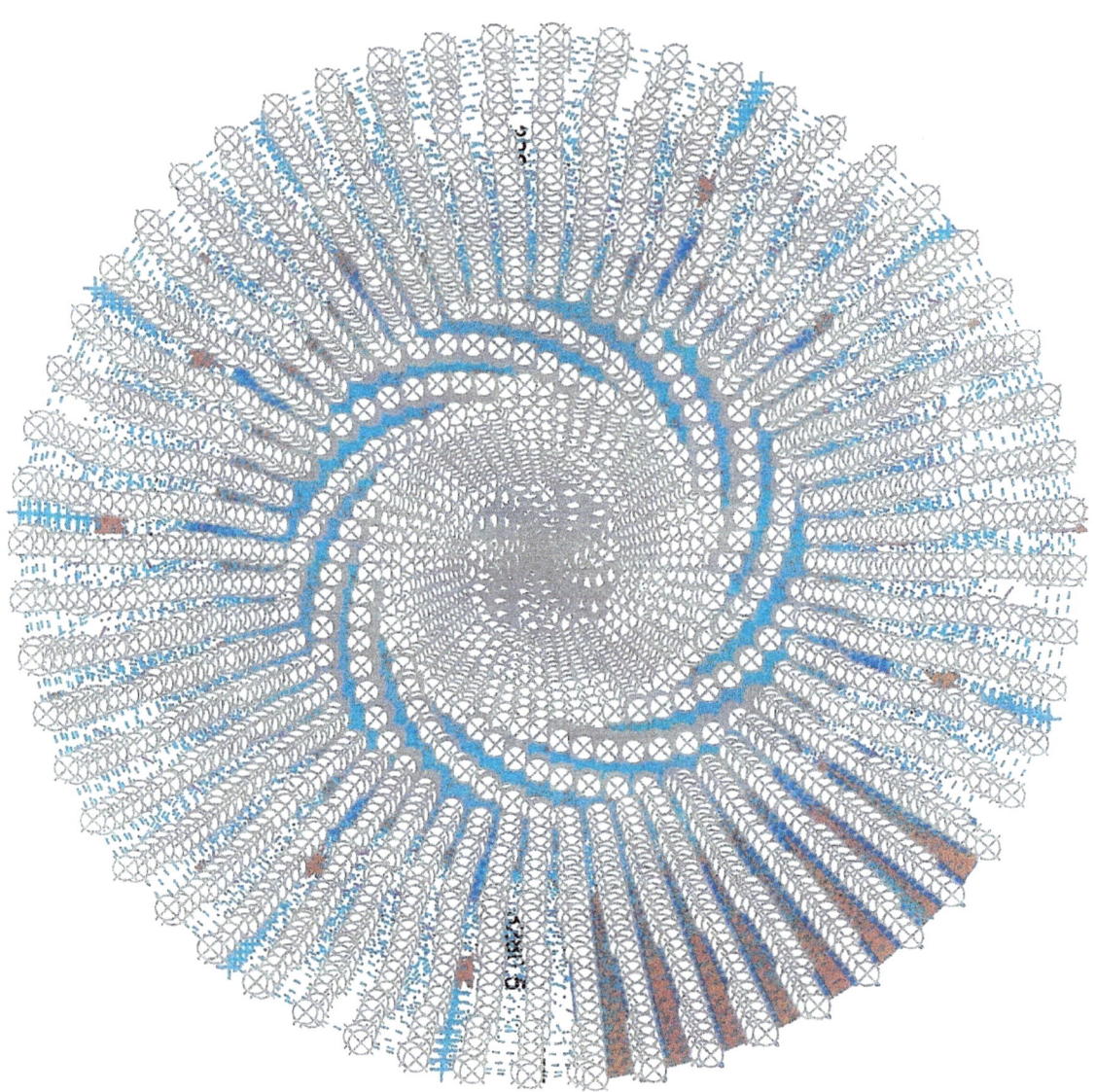

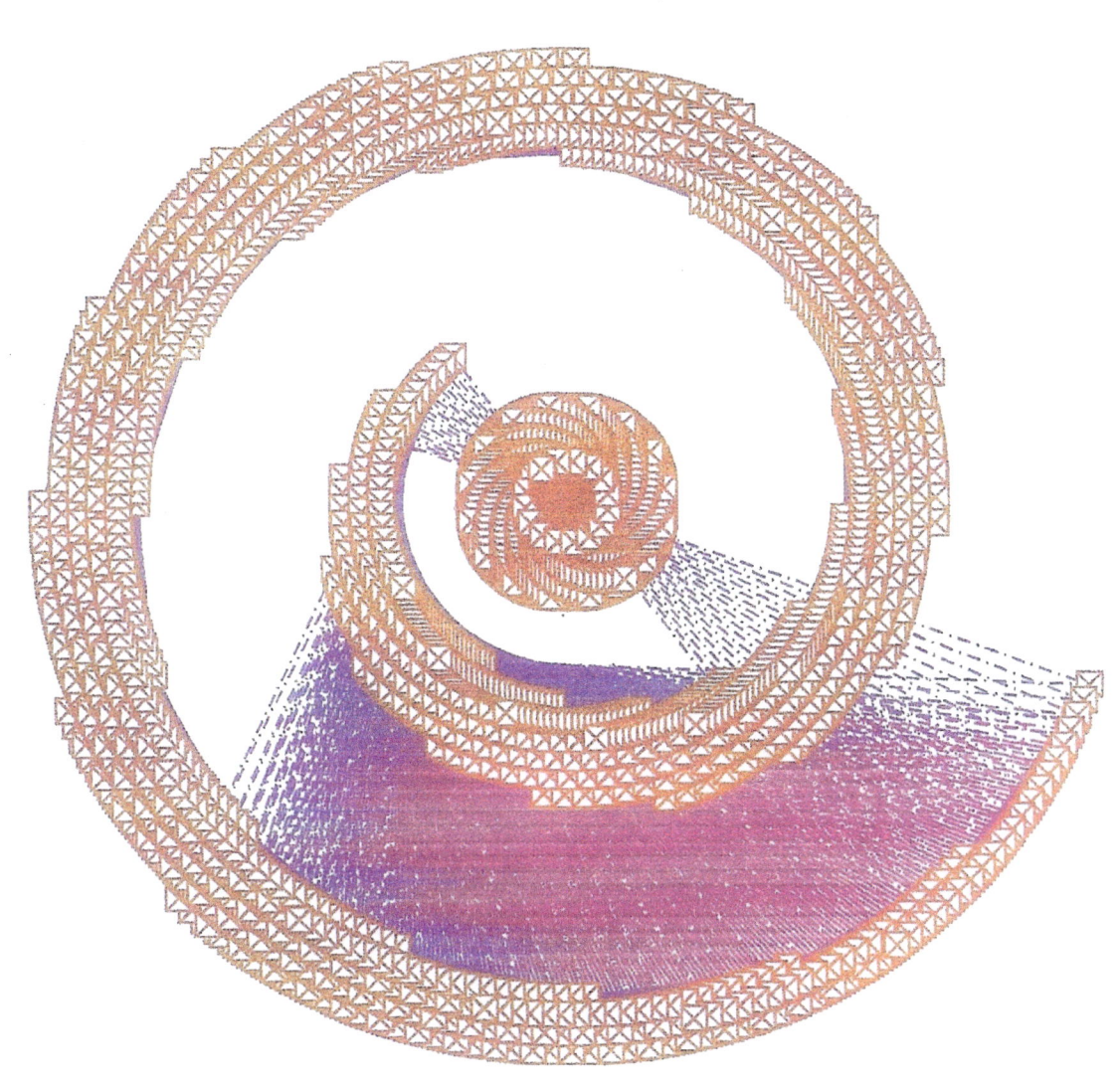

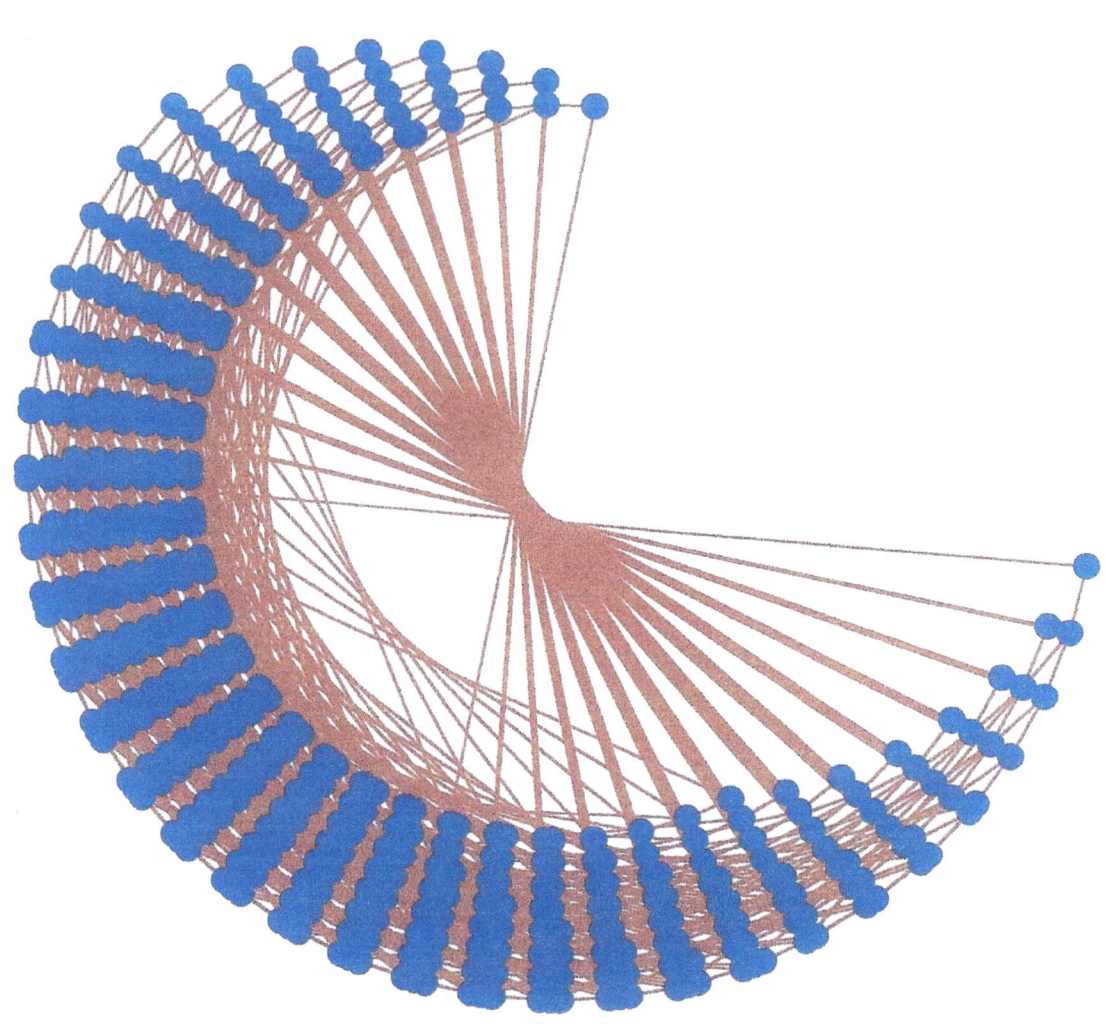

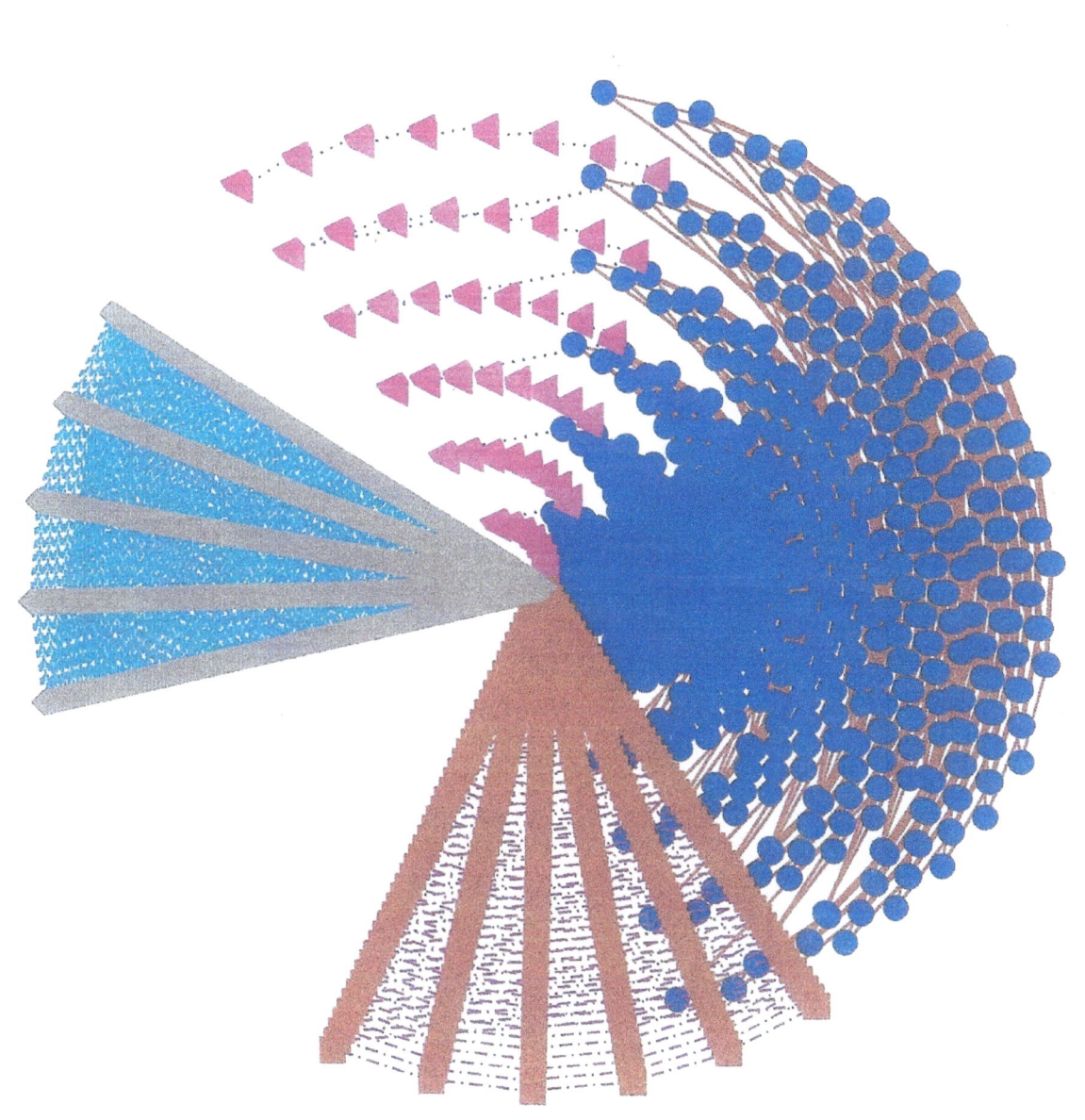

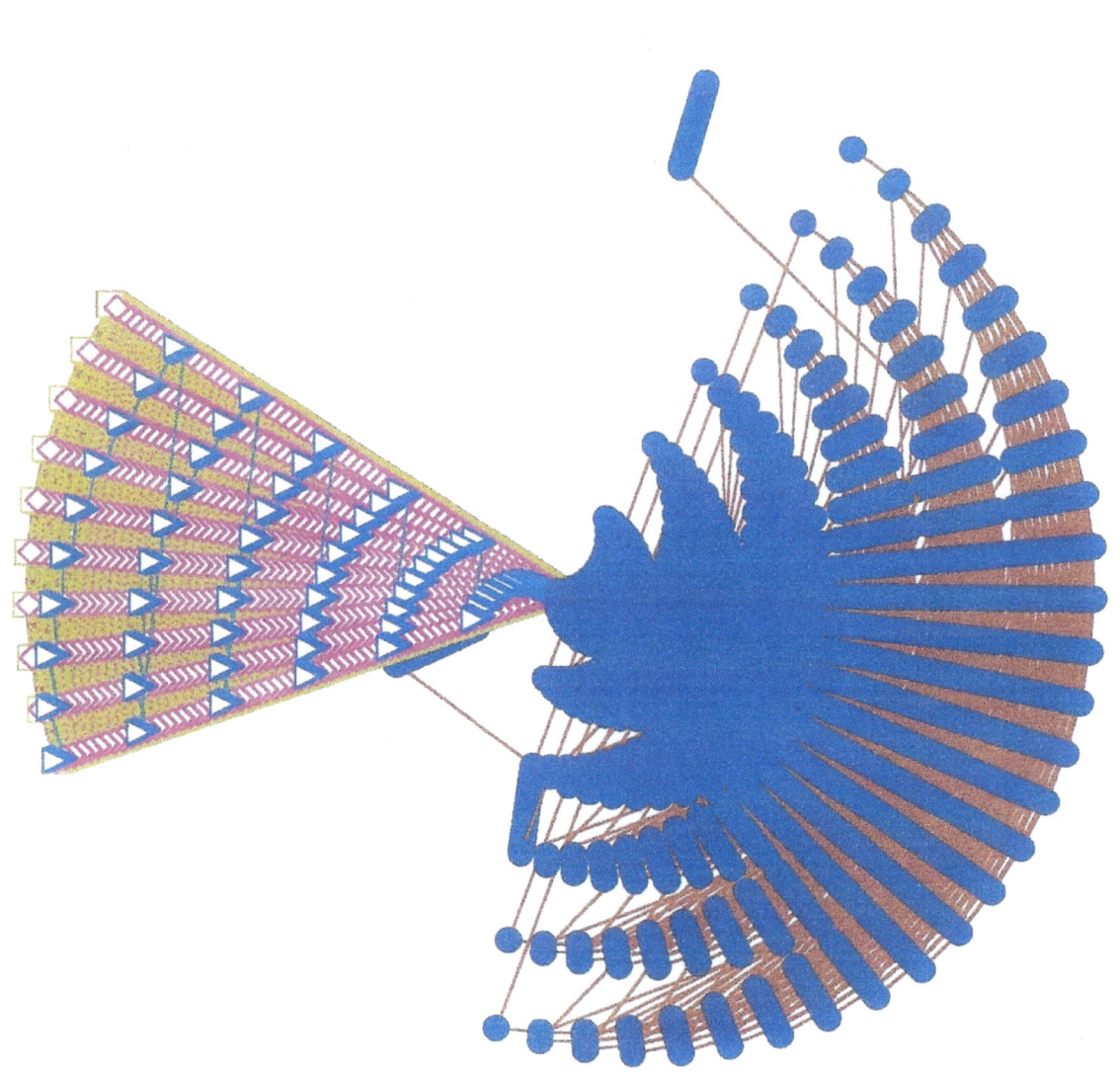

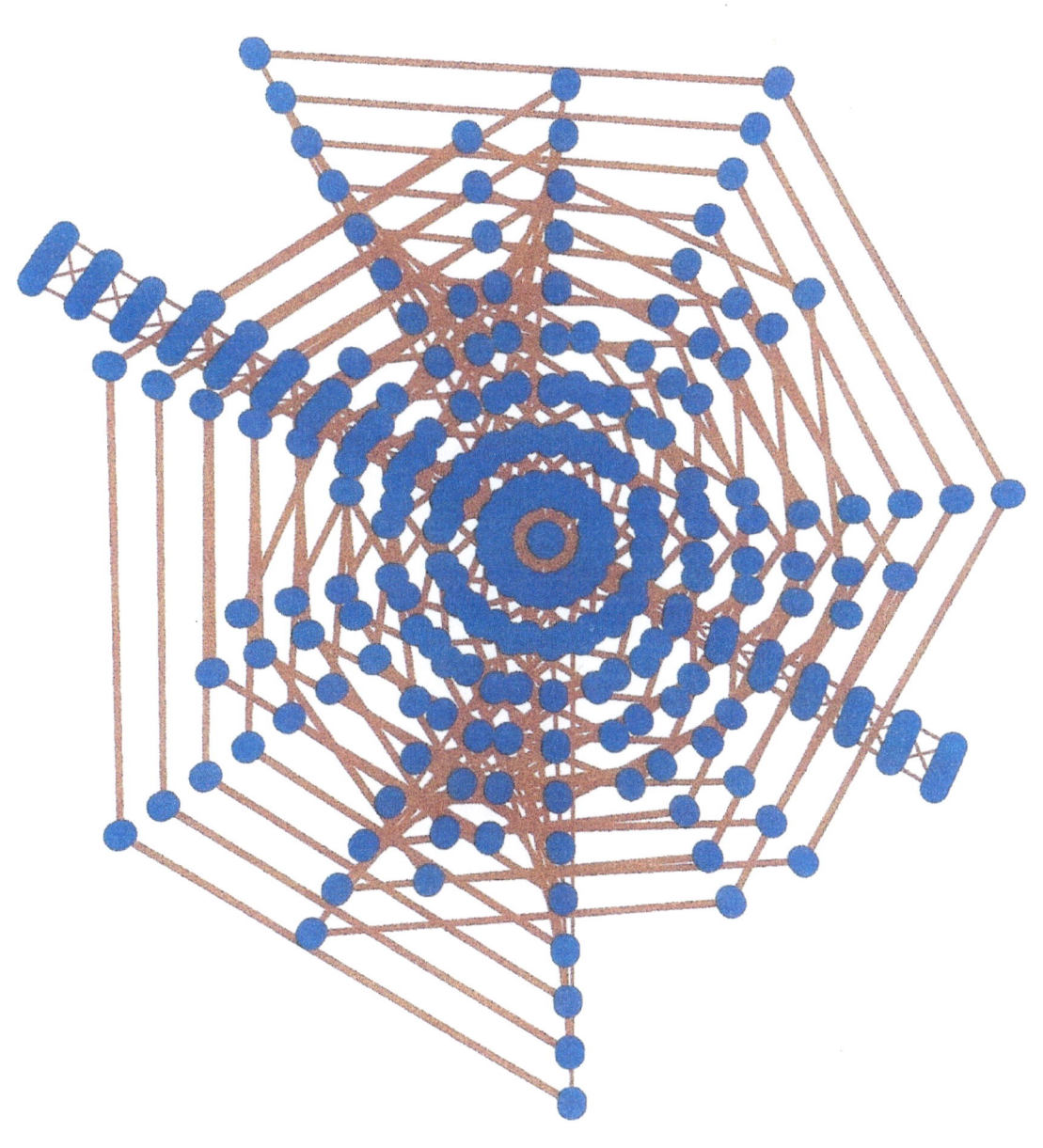

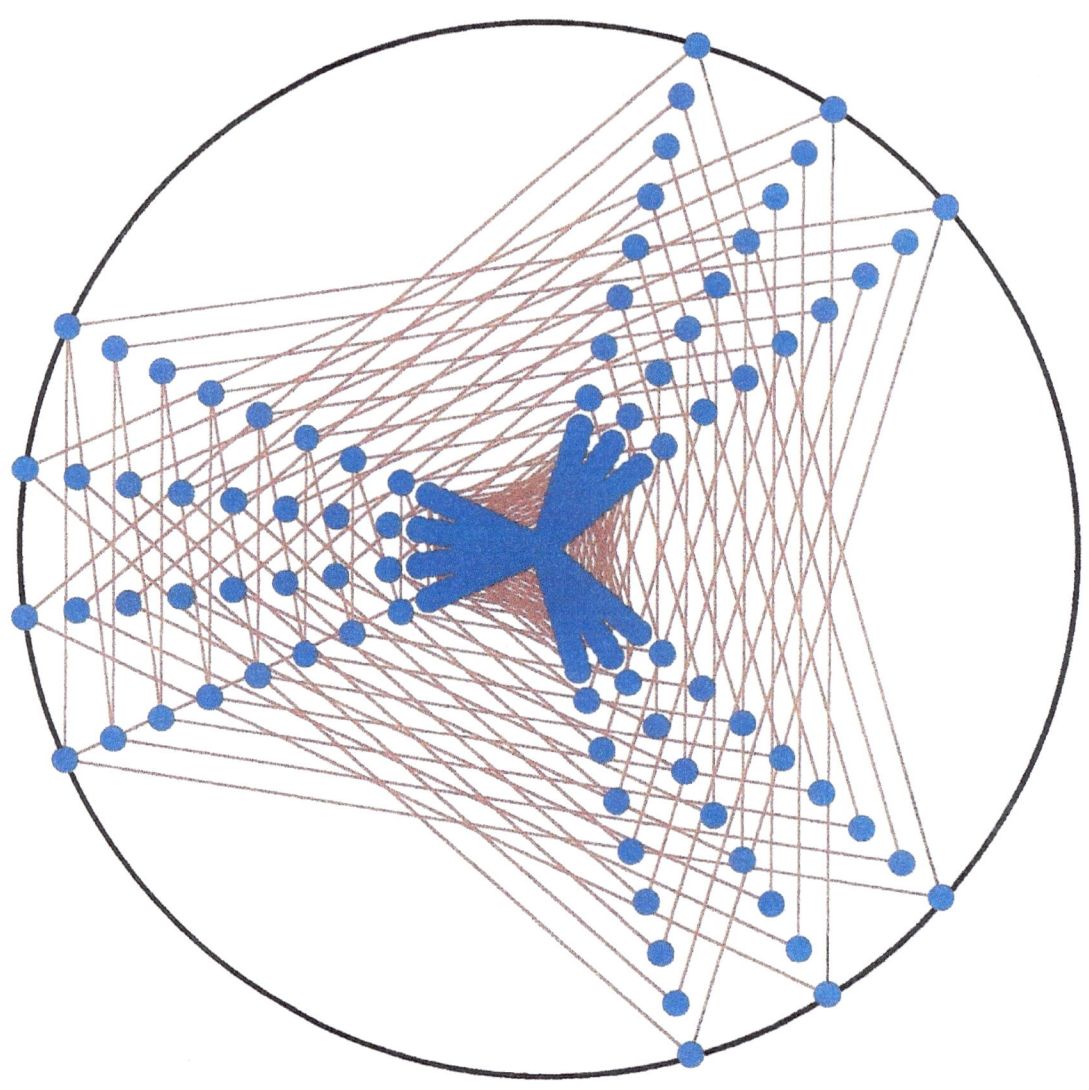

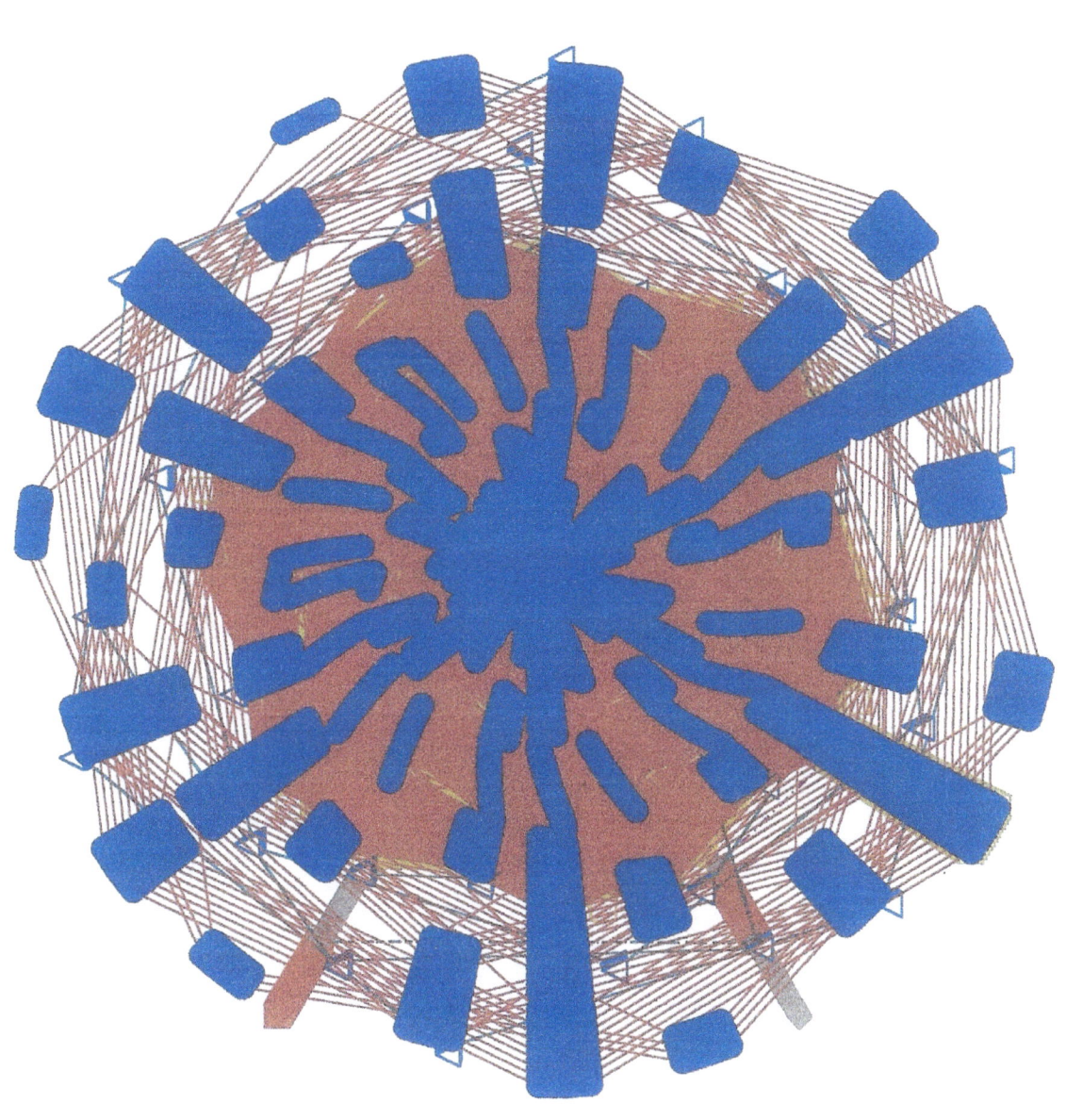

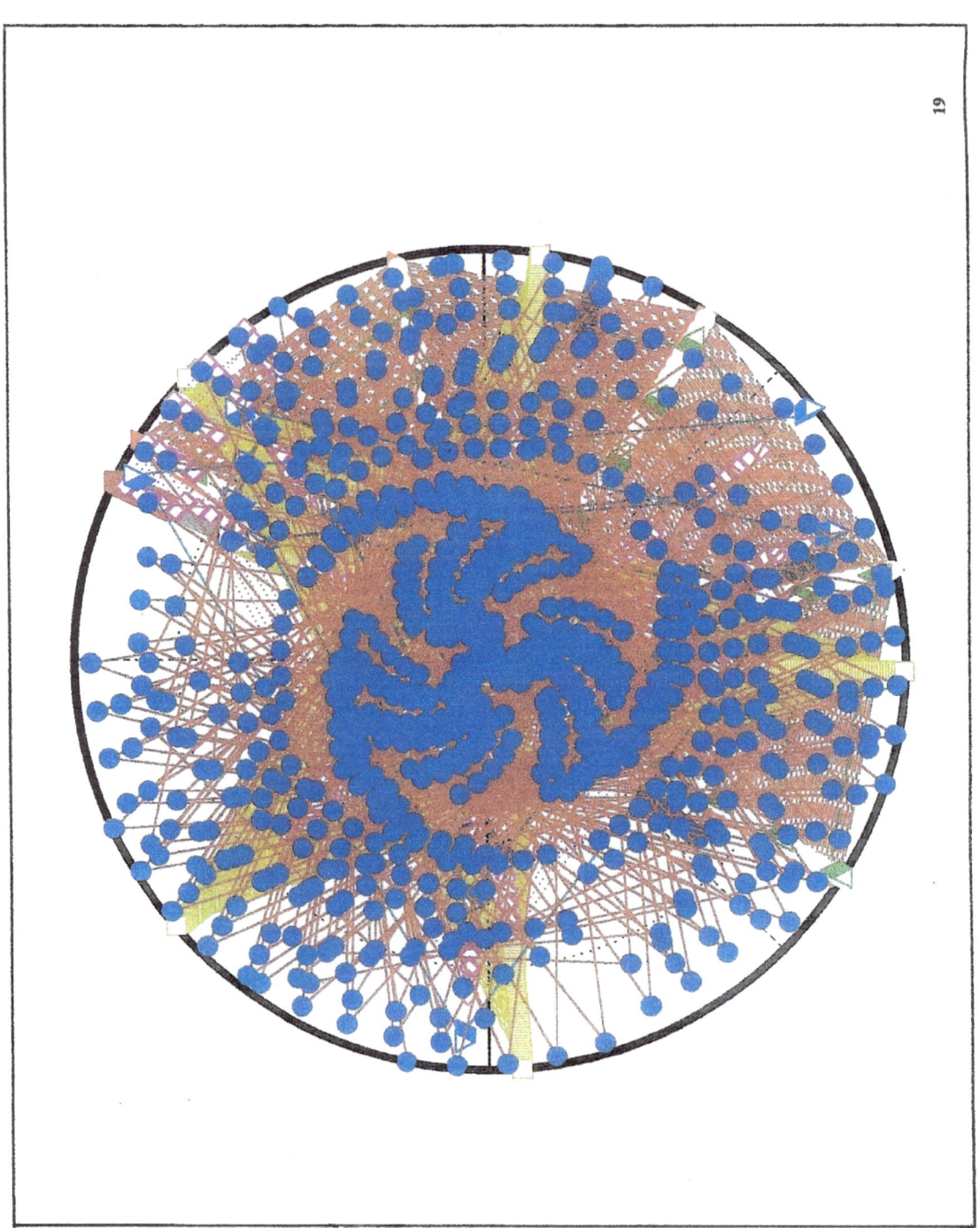

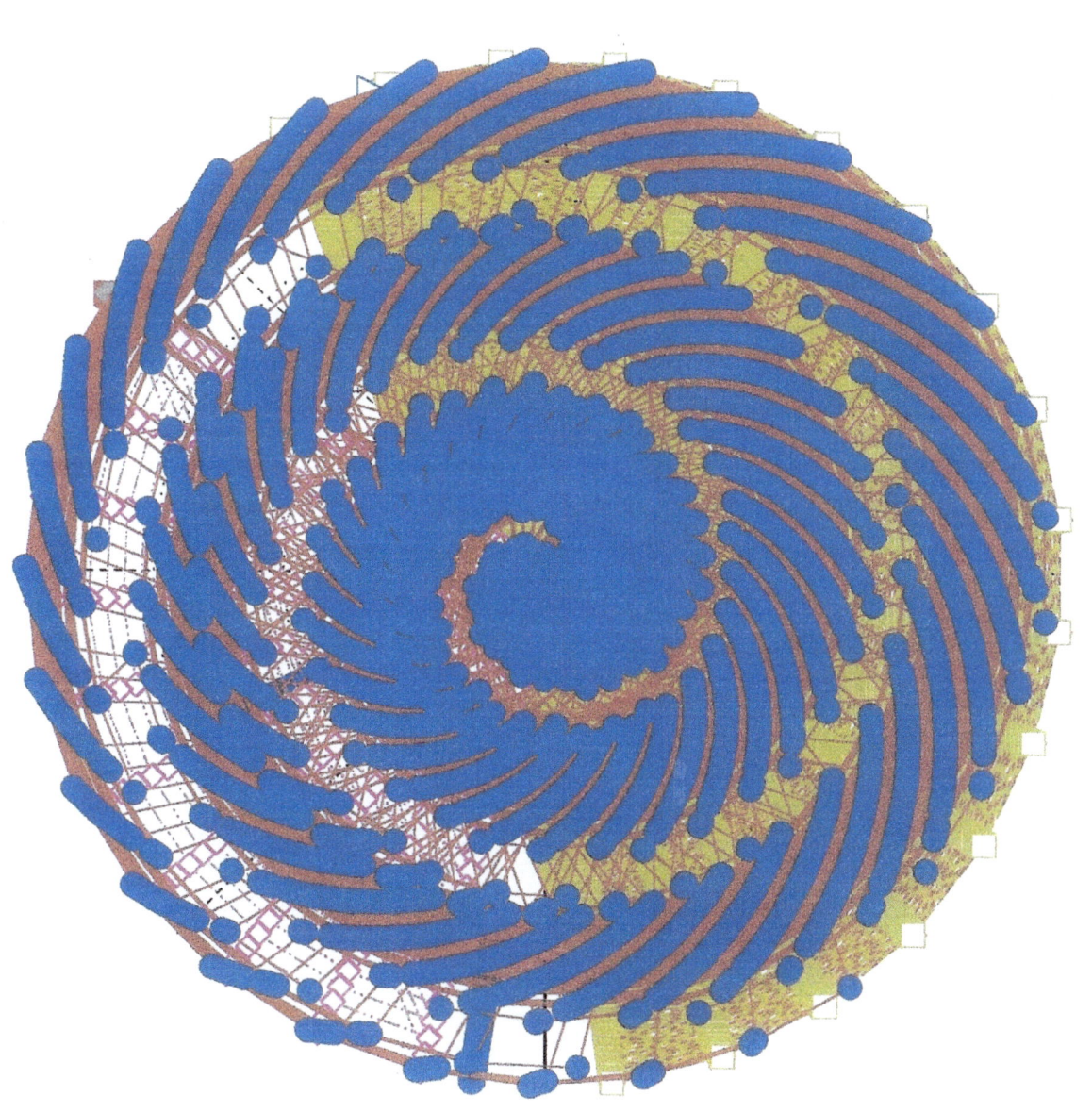

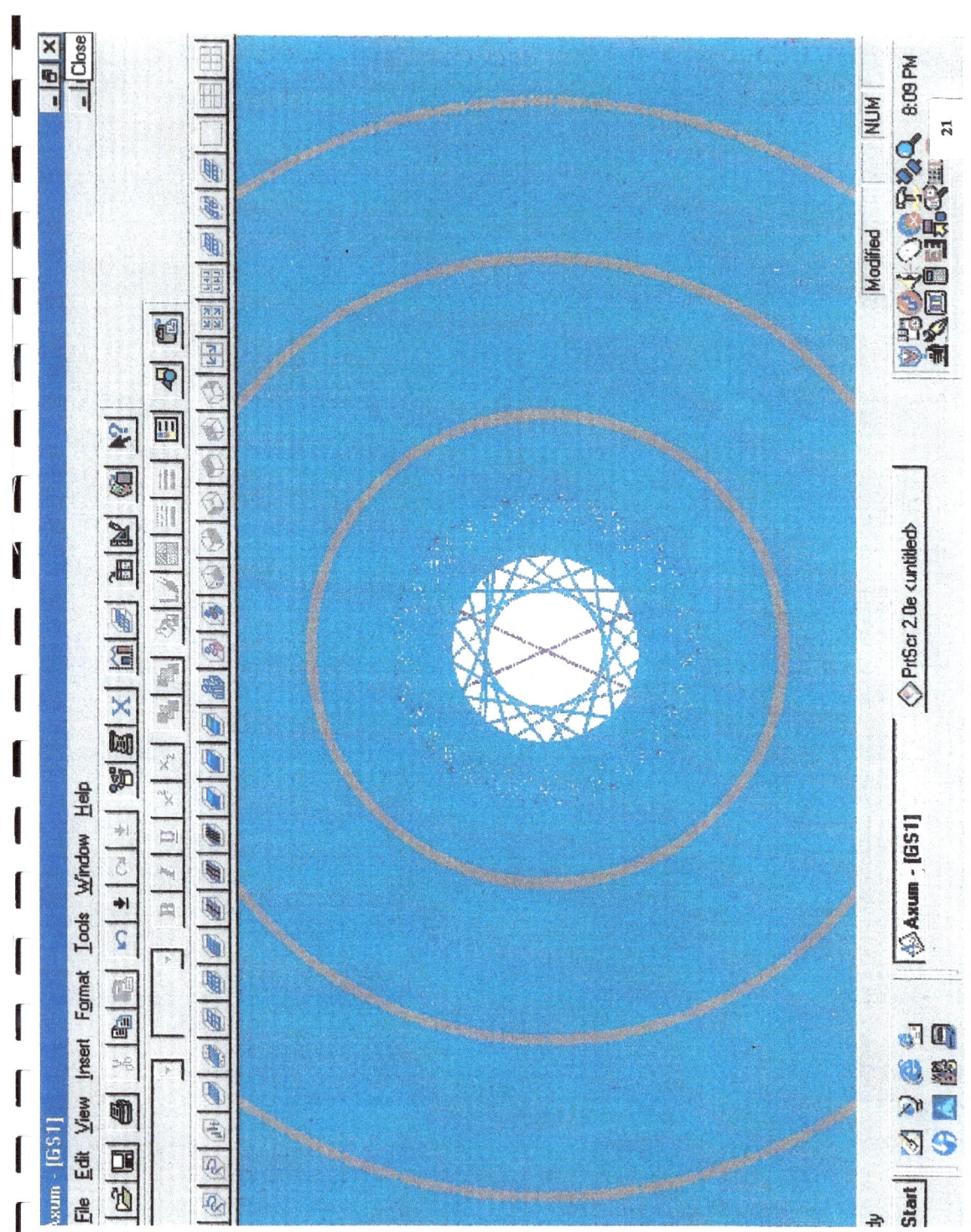

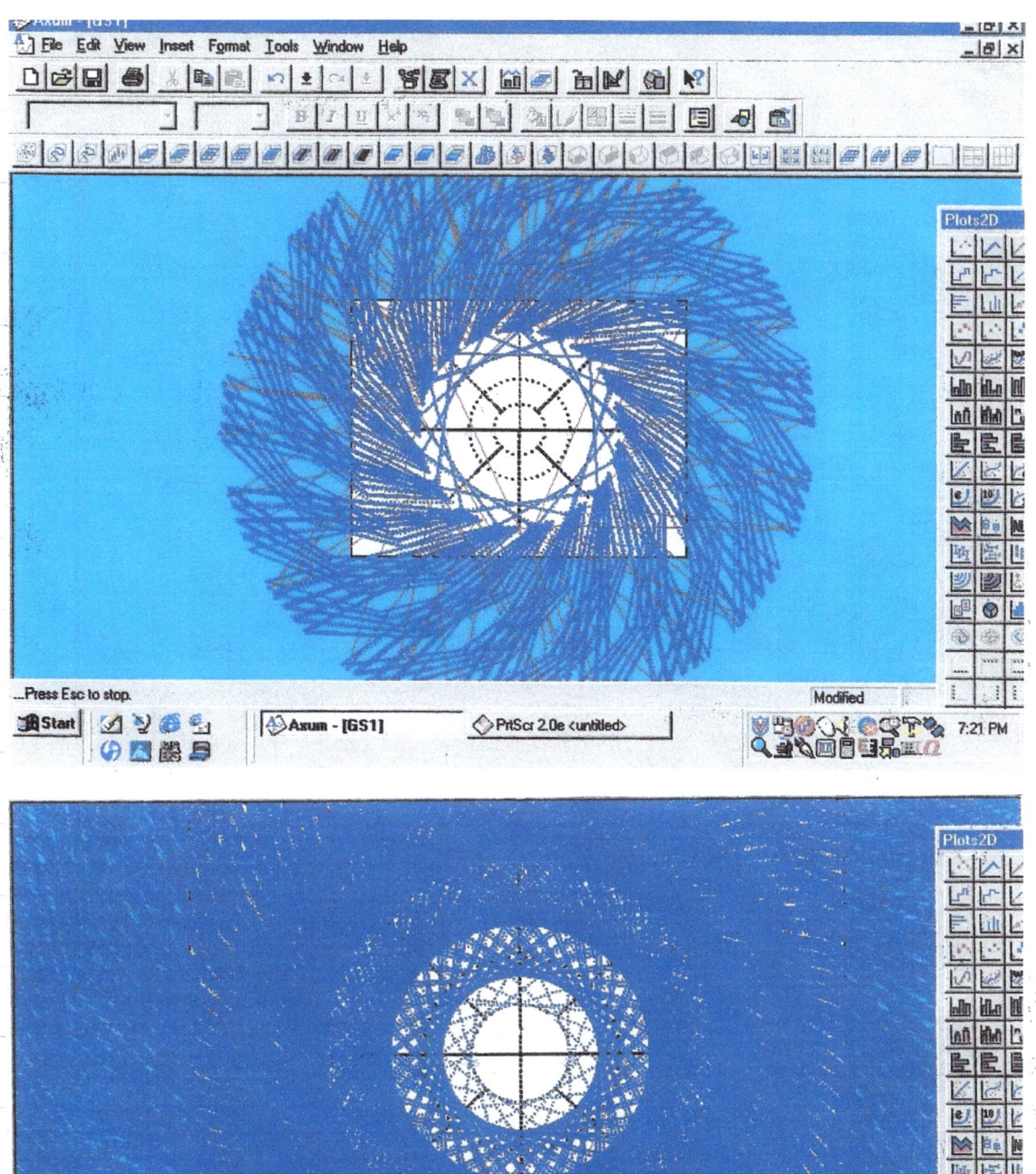

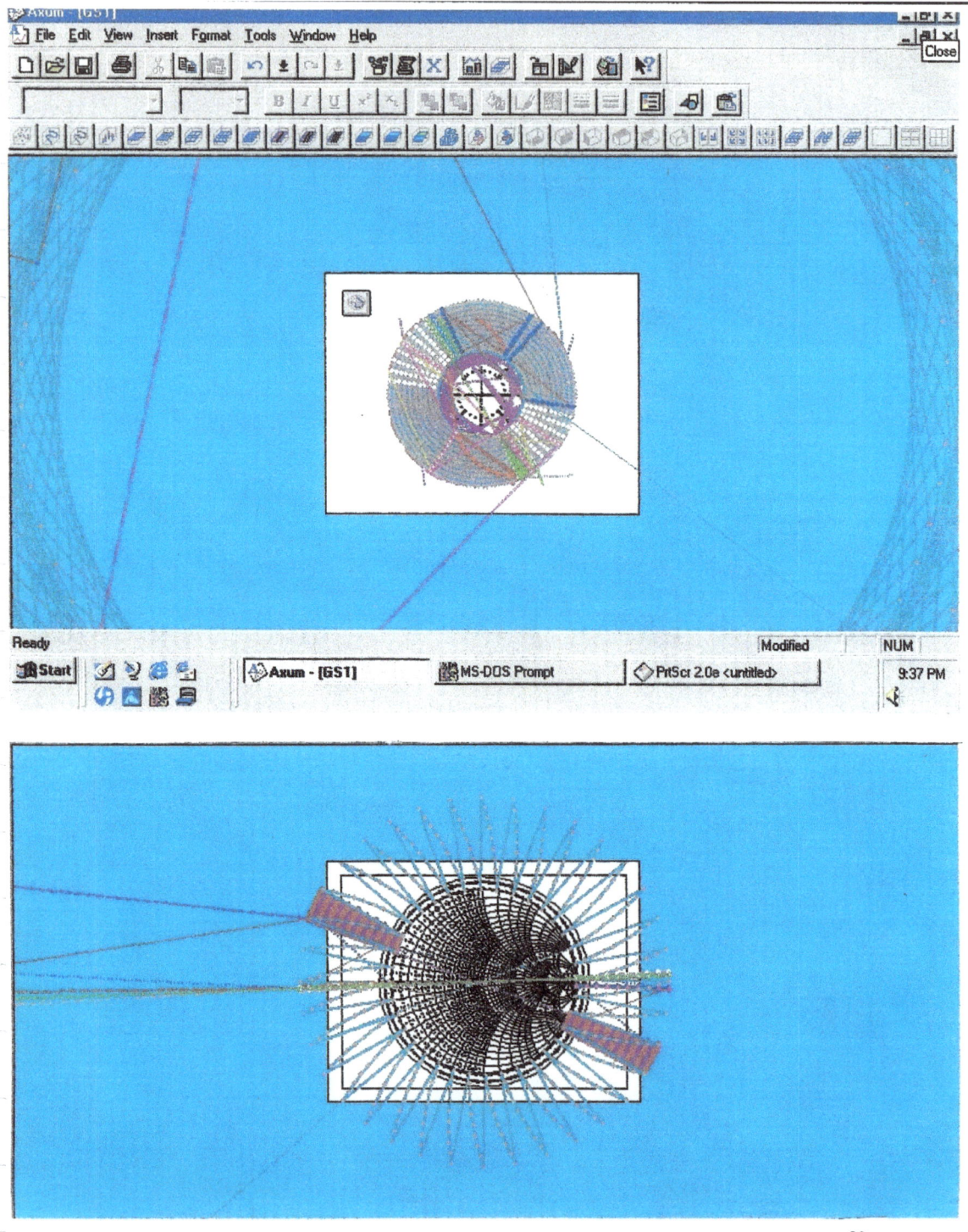

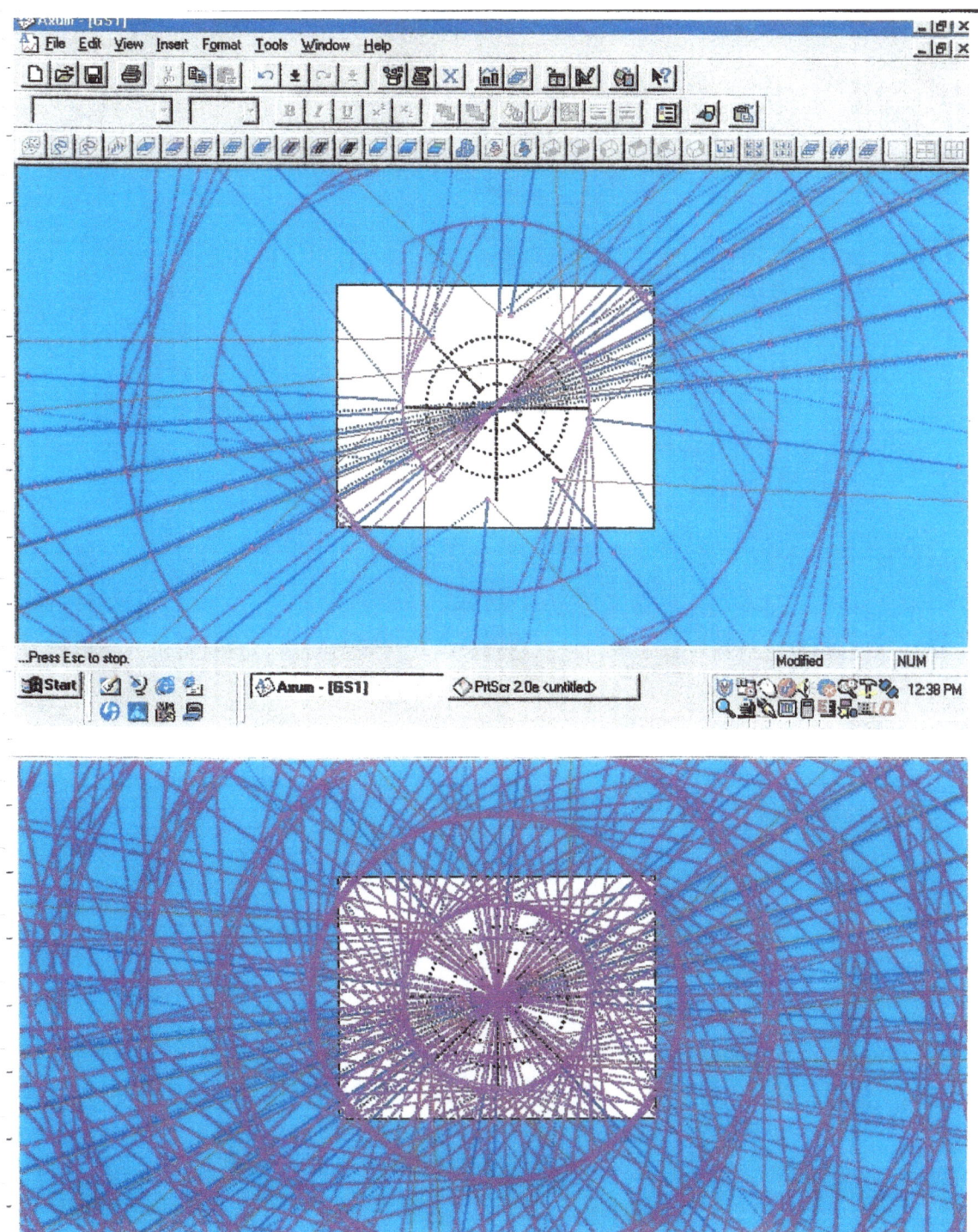

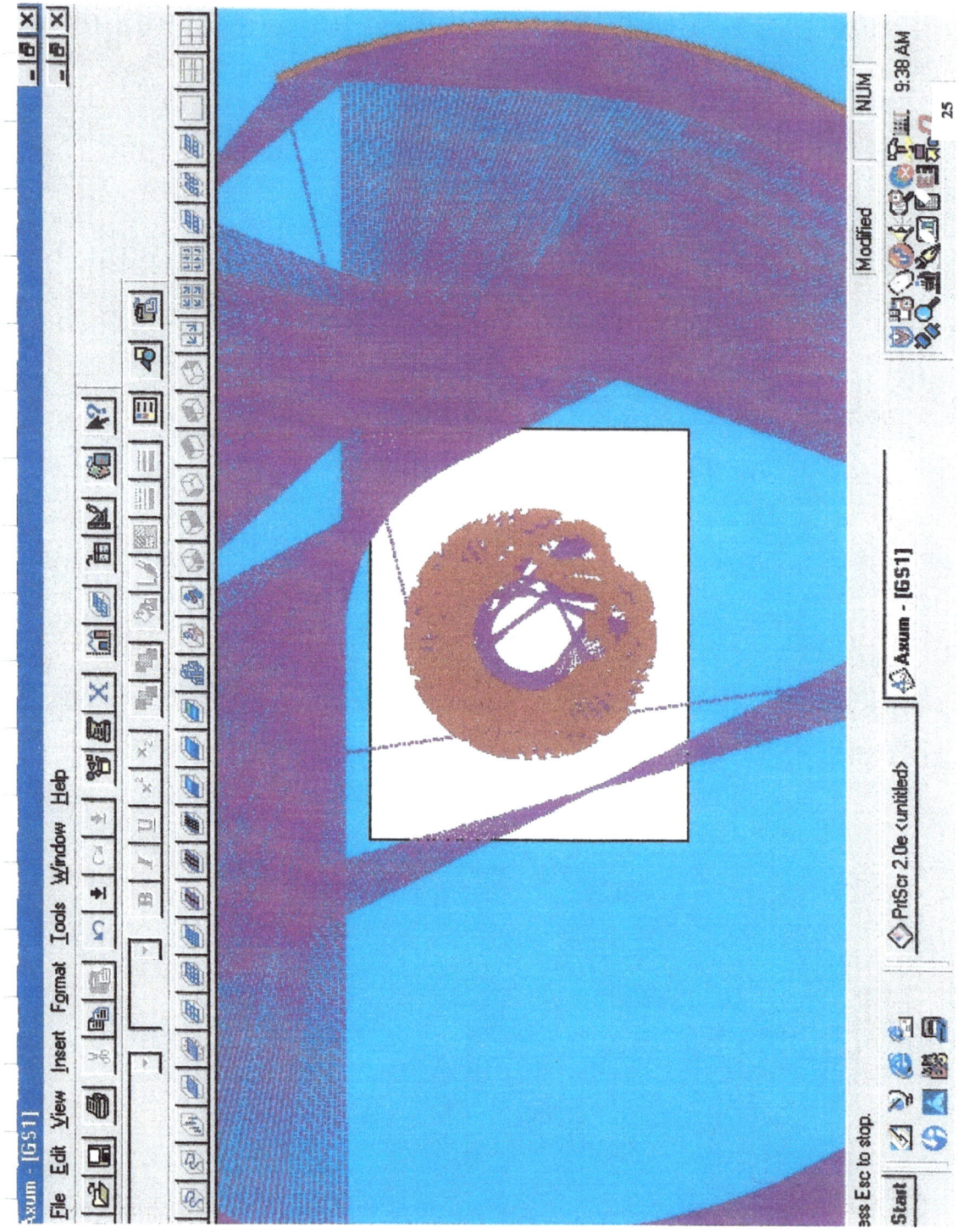

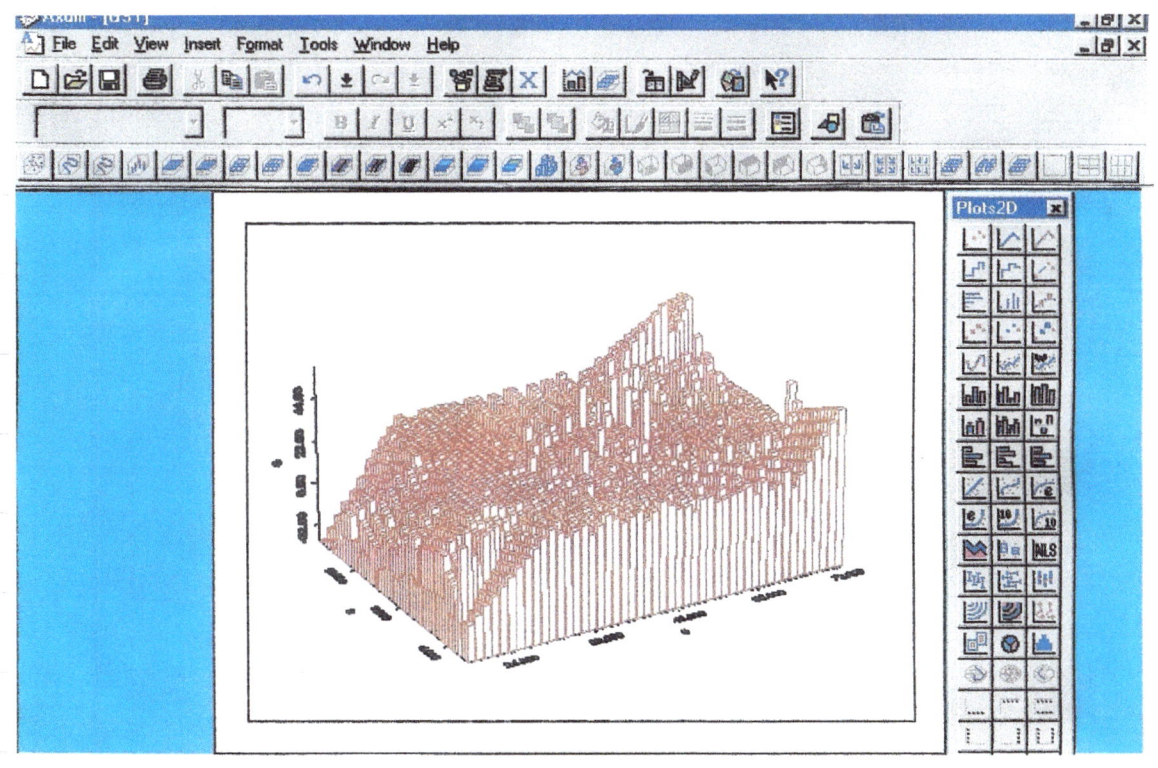

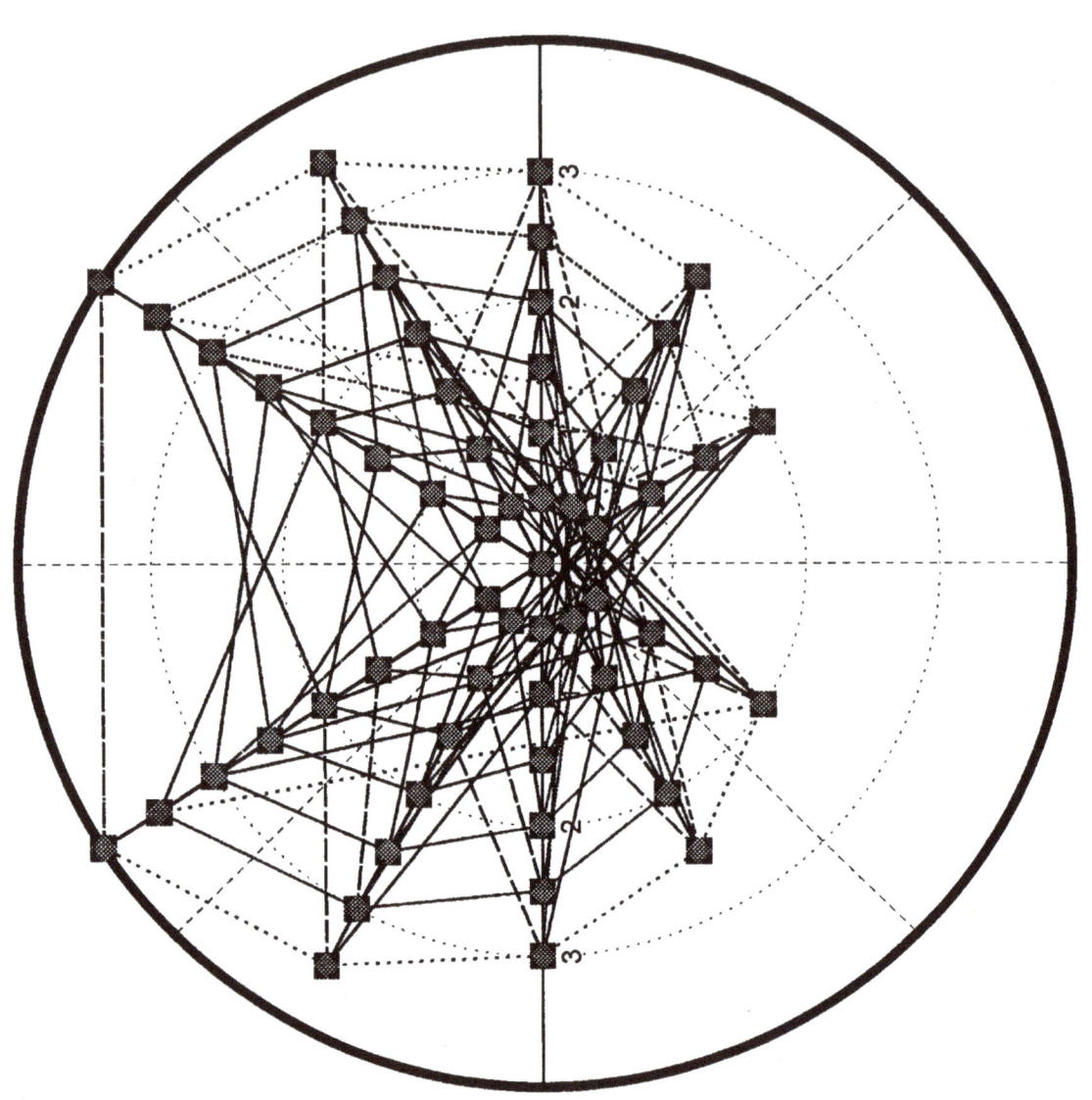

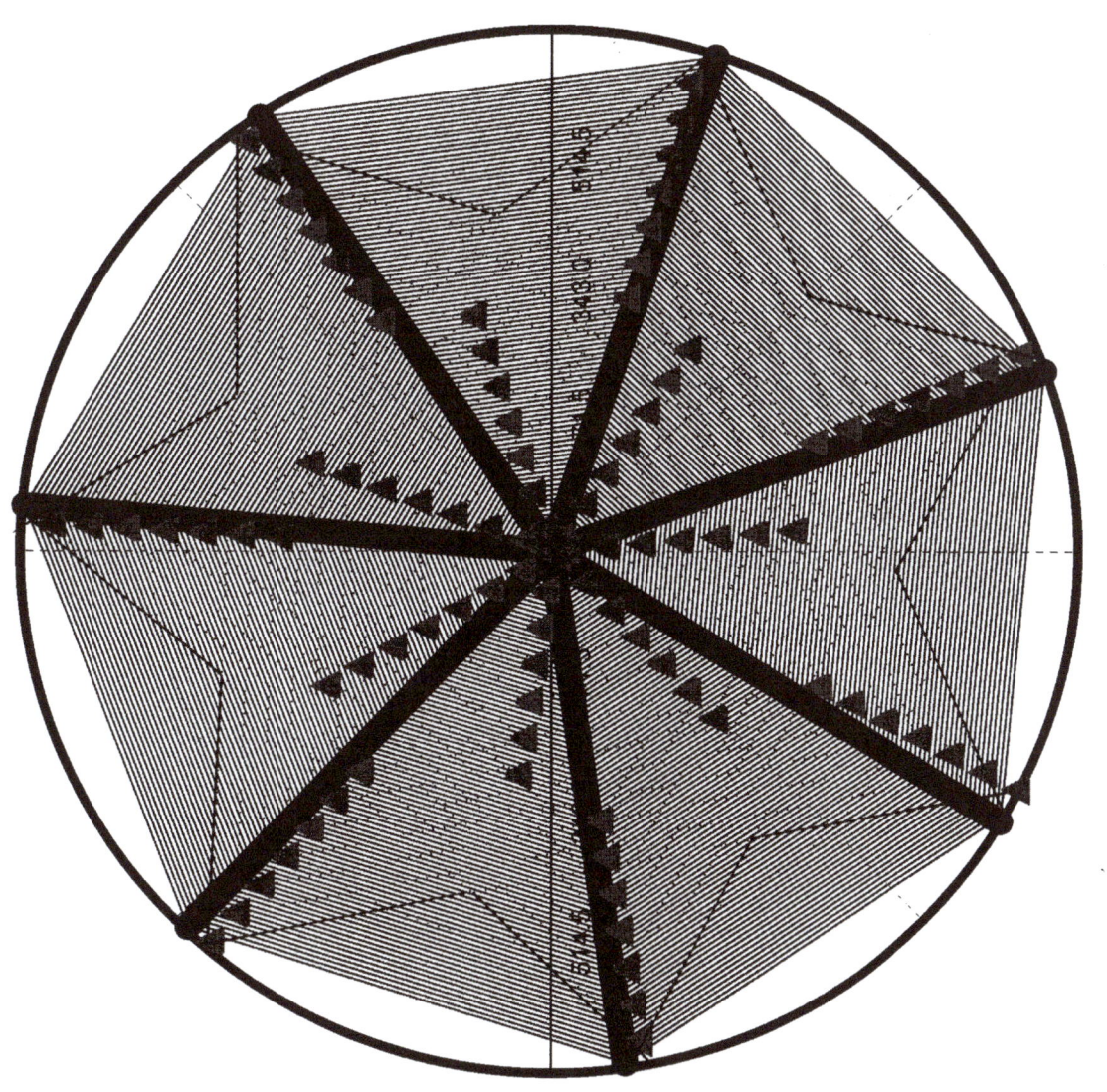

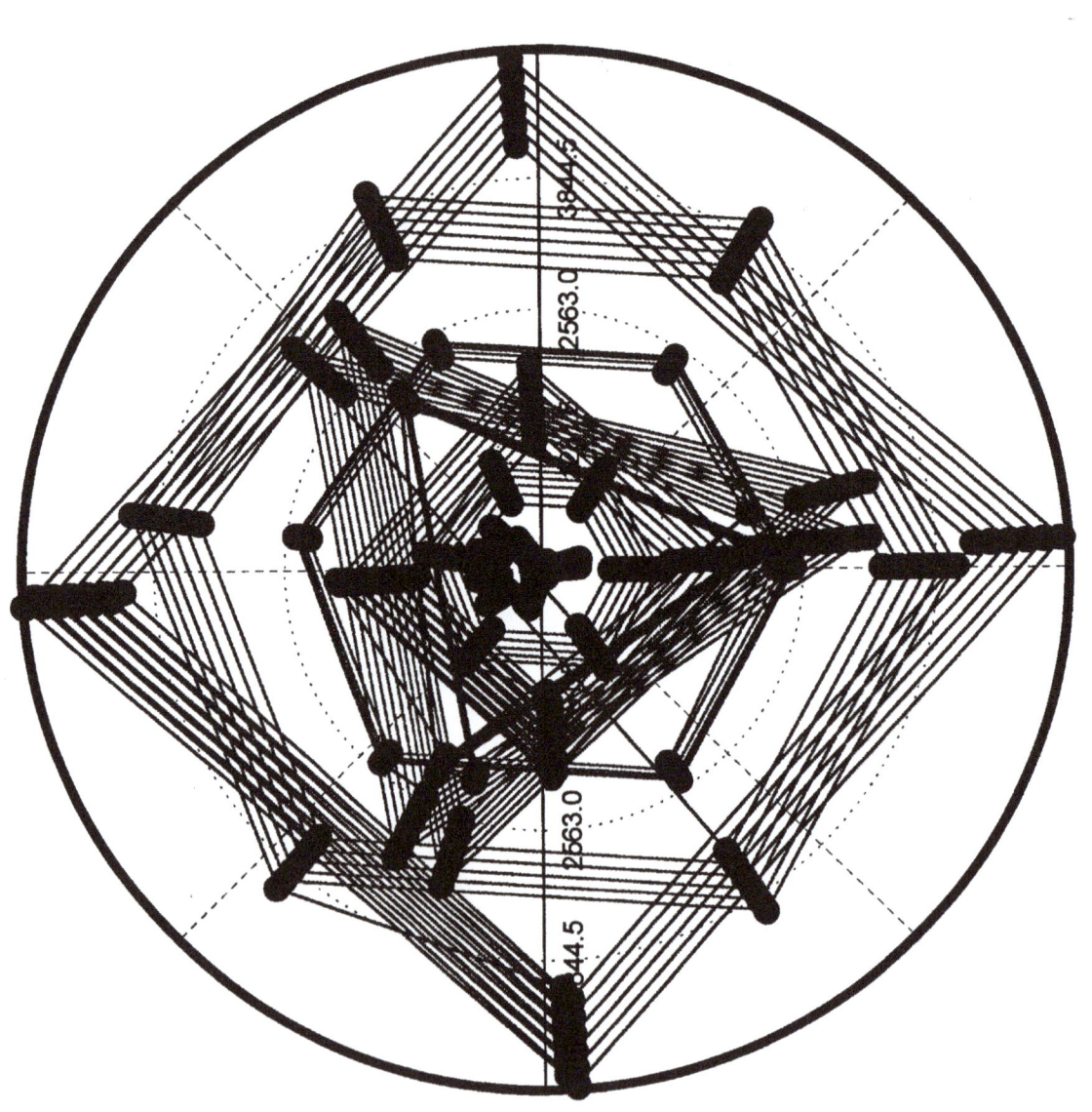

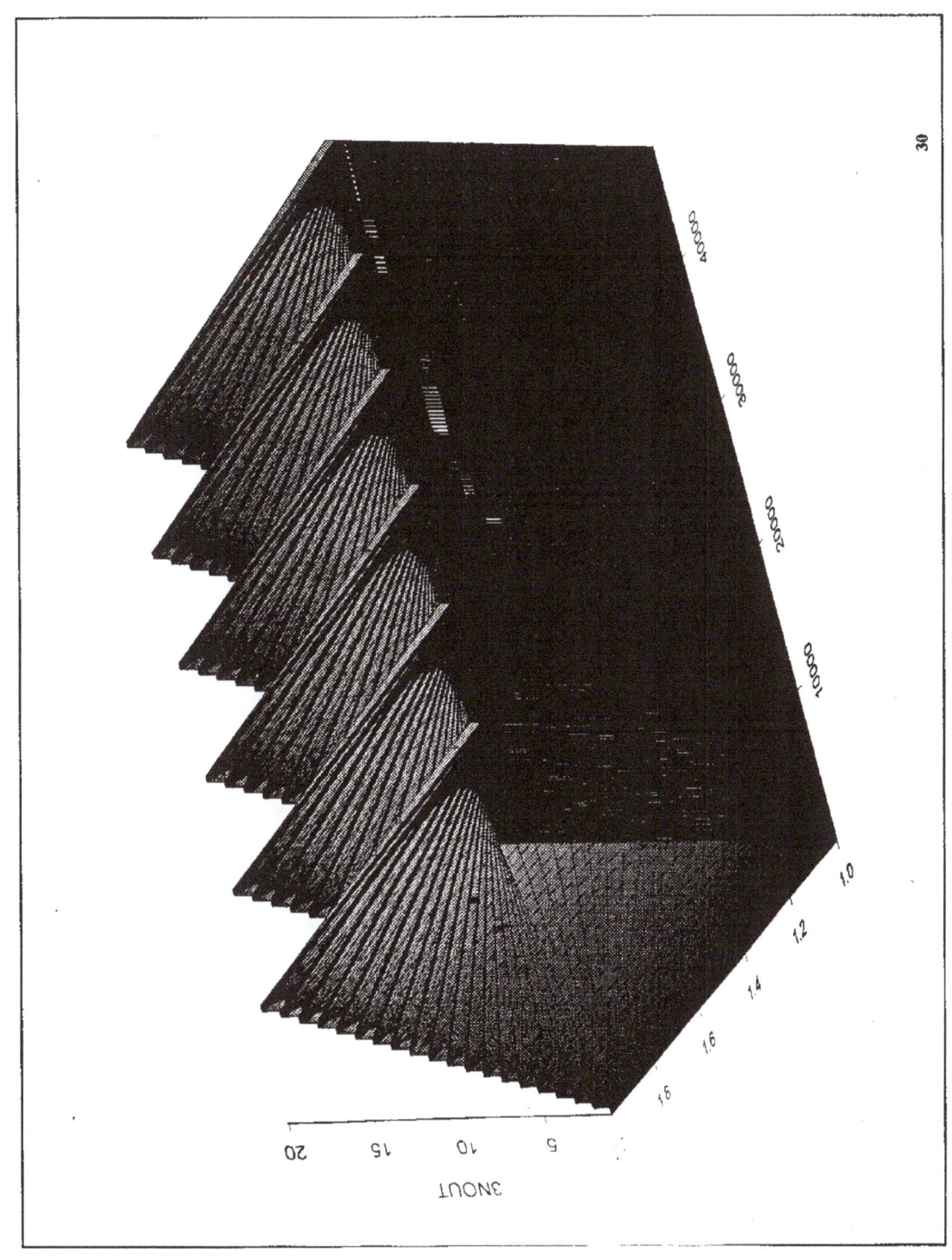

References

Mind design: philosophy, psychology, artificial intelligence. Edited by John Haugeland. Cambridge, Mass. : MIT Press, 1981

The Handbook of artificial intelligence. Edited by Avron Barr and Edward A. Feigenbaum. Reading, Mass. : Addison-Wesley, 1982

The science of artificial intelligence. Fred D'Ignazio & Allen L. Wold. New York: F. Watts, 1984

The mind's new science : a history of the cognitive revolution. Howard Gardner. New York: Basic Books, 1985

Artificial intelligence : theory, logic, and application. James F. Brule. Blue Rudge Summit, Pa. : TAB Books, 1986

What every engineer should know about artificial intelligence. William A. Taylor. Cambridge, Mass. : MIT Press, 1988

The expert executive : using AI and expert systems for financial management, marketing, production, and strategy. David Bendel Hertz. New York : Wiley, 1988

The emperor's new mind concerning computers, minds, and the laws of physics. Roger Penrose, Oxford; New York : Oxford University Press, 1989

Artificial experts : social knowledge and intelligent machines. H. M. Collins. Cambridge, Mass, : MIT Press, 1990

The creative mind: myths and mechanisms. Margaret A. Boden. New York, N.Y. : Basic Books, 1991

What computers still can't do : a critique of artificial reason. Hubert L. Dreyfus. Cambridge, Mass. : MIT Press, 1992

SOFTWARE

Microsoft Visual Basic

POLYN

Request free software, works with XP or WIN 7, at
gueller@aol.com

```
Private Sub cmdDISPLAY Click()
Cls
Print " Run...........Coefficient....................Data....................Product......
....................Sum           WORKING.....PLEASE WAIT UNTIL  < D O N E > is printed below"

AX = Val(B1.Text)
BX = Val(B2.Text)
CX = Val(B3.Text)
DX = Val(B4.Text)
EX = Val(B5.Text)
FX = Val(B6.Text)
GX = Val(B7.Text)
AY = Val(B8.Text)
Q = Val(Text1.Text)
L(1) = Val(A1.Text)
L(2) = Val(A2.Text)
L(3) = Val(A3.Text)

L(4) = Val(A4.Text)
L(5) = Val(A5.Text)
L(6) = Val(A6.Text)
L(7) = Val(A7.Text)
L(8) = Val(A8.Text)

Open "C:\POLYN\POLYNC.TXT" For Output As #1
H = 1
For I1 = 1 To 8
For I2 = 1 To 8
If I2 = I1 Then GoTo 99
For I3 = 1 To 8
If I3 = I1 Or I3 = I2 Then GoTo 98
For I4 = 1 To 8
If I4 = I1 Or I4 = I2 Or I4 = I3 Then GoTo 97
For I5 = 1 To 8
If I5 = I1 Or I5 = I2 Or I5 = I3 Or I5 = I4 Then GoTo 96
For I6 = 1 To 8
If I6 = I1 Or I6 = I2 Or I6 = I3 Or I6 = I4 Or I6 = I5 Then GoTo 95
For I7 = 1 To 8
If I7 = I1 Or I7 = I2 Or I7 = I3 Or I7 = I4 Or I7 = I5 Or I7 = I6 Then GoTo 94
Let I8 = 36 - (I1 + I2 + I3 + I4 + I5 + I6 + I7)

K = L(I1) * AX
LL = L(I2) * BX
M = L(I3) * CX
N = L(I4) * DX
O = L(I5) * EX
P = L(I6) * FX
R = L(I7) * GX
KK = L(I8) * AY

TT(H) = K + LL + M + N + O + P + R + KK

If H < 25 Then GoTo 11 Else GoTo 12
11 Print " "; H; ": "; AX; BX; CX; DX; EX; FX; GX; AY; "     "; L(I1); L(I2); L(I3); L(I4); L(I
5); L(I6); L(I7); L(I8); "     "; K; LL; M; N; O; P; R; KK; "     "; TT(H)
12 Write #1, H, AX, BX, CX, DX, EX, FX, GX, AY, L(I1), L(I2), L(I3), L(I4), L(I5), L(I6), L(I7),
 L(I8), K, LL, M, N, O, P, R, KK, TT(H)
If H = Q Then GoTo 4

H = H + 1
94 Next I7
95 Next I6
96 Next I5
97 Next I4
98 Next I3
99 Next I2
   Next I1
4 Close #1
```

```
Form1 - 2

Print "                        ---------- DONE ----------
                                   * Use a grapher to analize the output, find the file at C:\PO
LYN\POLYN.DAT *"   .
End·Sub

Private Sub cmdDISPLAYR Click()
Cls
Print " Run.................Range.....................Data........................Product......
.....................Sum          WORKING.....PLEASE WAIT UNTIL  < D O N E > is printed below"

C10 = Val(C10.Text)
C12 = Val(C12.Text)
C1 = Val(C1.Text)
C20 = Val(C20.Text)
C22 = Val(C22.Text)
C2 = Val(C2.Text)
C30 = Val(C30.Text)
C32 = Val(C32.Text)
C3 = Val(C3.Text)
C40 = Val(C40.Text)
C42 = Val(C42.Text)
C4 = Val(C4.Text)
C50 = Val(C50.Text)
C52 = Val(C52.Text)
C5 = Val(C5.Text)
C60 = Val(C60.Text)
C62 = Val(C62.Text)
C6 = Val(C6.Text)
C70 = Val(C70.Text)
C72 = Val(C72.Text)
C7 = Val(C7.Text)
C80 = Val(C80.Text)
C82 = Val(C82.Text)
C8 = Val(C8.Text)

QQ = Val(Text4.Text)
XX66 = Val(Text2.Text)
YY77 = Val(Text3.Text)

L(1) = Val(A1.Text)
L(2) = Val(A2.Text)
L(3) = Val(A3.Text)
L(4) = Val(A4.Text)
L(5) = Val(A5.Text)
L(6) = Val(A6.Text)
L(7) = Val(A7.Text)
L(8) = Val(A8.Text)

Open "C:\POLYN\POLYNR.TXT" For Output As #2

H = 1
For I1 = 1 To 8
For I2 = 1 To 8
If I2 = I1 Then GoTo 199
For I3 = 1 To 8
If I3 = I1 Or I3 = I2 Then GoTo 198
For I4 = 1 To 8
If I4 = I1 Or I4 = I2 Or I4 = I3 Then GoTo 197
For I5 = 1 To 8
If I5 = I1 Or I5 = I2 Or I5 = I3 Or I5 = I4 Then GoTo 196
For I6 = 1 To 8
If I6 = I1 Or I6 = I2 Or I6 = I3 Or I6 = I4 Or I6 = I5 Then GoTo 195
For I7 = 1 To 8
If I7 = I1 Or I7 = I2 Or I7 = I3 Or I7 = I4 Or I7 = I5 Or I7 = I6 Then GoTo 194
Let I8 = 36 - (I1 + I2 + I3 + I4 + I5 + I6 + I7)

For YY = C80 To C82 Step C8
For Y = C70 To C72 Step C7
```

130

```
·For XXXXXX = C60 To C62 Step C6
 For XXXXX = C50 To C52 Step C5
 For XXXX = C40 To C42 Step C4
 For XXX = C30 To C32 Step C3
 For XX = C20 To C22 Step C2
 For X = C10 To C12 Step C1

 Z1 = (L(I1) * X)
 Z2 = (L(I2) * XX)
 Z3 = (L(I3) * XXX)
 Z4 = (L(I4) * XXXX)
 Z5 = (L(I5) * XXXXX)
 Z6 = (L(I6) * XXXXXX)
 Z7 = (L(I7) * Y)
 Z8 = (L(I8) * YY)

 A(H) = Z1 + Z2 + Z3 + Z4 + Z5 + Z6 + Z7 + Z8
 If A(H) > XX66 Or A(H) < YY77 Then GoTo 217

 If H = (QQ + 1) Then GoTo 777

 If H < 25 Then GoTo 11 Else GoTo 12
 11 Print " "; H; ; "     "; X; XX; XXX; XXXX; XXXXX; XXXXXX; Y; YY; "    "; L(I1); L(I2); L(I3);
   L(I4); L(I5); L(I6); L(I7); L(I8); "    "; Z1; Z2; Z3; Z4; Z5; Z6; Z7; Z8; "   "; A(H)
 12 Write #2, H, X, XX, XXX, XXXX, XXXXX, XXXXXX, Y, YY, L(I1), L(I2), L(I3), L(I4), L(I5), L(I6)
 , L(I7), L(I8), Z1, Z2, Z3, Z4, Z5, Z6, Z7, Z8, A(H)

 H = H + 1

 217 Next X
     Next XX
     Next XXX
     Next XXXX
     Next XXXXX
     Next XXXXXX
     Next Y
     Next YY
 194 Next I7
 195 Next I6
 196 Next I5
 197 Next I4
 198 Next I3
 199 Next I2
     Next I1
 777 Close #2
 Print "                       ---------- DONE ----------
                         * Use a grapher to analize the output, find the file at C:\PO
 LYN\POLYN.DAT *"
 End Sub

 Private Sub cmdExit Click()
 Unload Me
 End
 End Sub

 Private Sub Label10 Click()

 End Sub

 Private Sub Label12 Click()

 End Sub

 Private Sub Label11 Click()

 End Sub

 Private Sub Label13 Click()
```

Form1 - 4

End Sub

Private Sub Label15 Click()

End Sub

Private Sub Label7 Click()

End Sub

Module1 - 1

```
Option Base 1
Public L(100) As Single
Public A(300000) As Single
Public TT(300000) As Single
Public H As Integer
Public K As Single
Public LL As Single
Public M As Single
Public N As Single
Public O As Single
Public P As Single
Public R As Single
Public KK As Single
Public Q As Integer
Public QQ As Integer
Public Z1 As Single
Public Z2 As Single
Public Z3 As Single
Public Z4 As Single
Public Z5 As Single
Public Z6 As Single
Public Z7 As Single
Public Z8 As Single
```

SOFTWARE

INFER

Microsoft Visual Basic

Request free software, works with XP or WIN 7, at
gueller@aol.com

```
Form1 - 1

Private Sub cmdDISPLAY_Click()
Cls
Print "                                                                        WORK
ING.....PLEASE WAIT UNTIL < D O N E > is printed at the bottom"
Open "C:\INFER\INFER.TXT" For Output As #1
I = 1
For I1 = 1 To 10 Step 1
For I2 = 1 To 10 Step 1
If I2 = I1 Then GoTo 108
For I3 = 1 To 10 Step 1
If I3 = I1 Or I3 = I2 Then GoTo 107
For I4 = 1 To 10 Step 1
If I4 = I1 Or I4 = I2 Or I4 = I3 Then GoTo 106
For I5 = 1 To 10 Step 1
If I5 = I1 Or I5 = I2 Or I5 = I3 Or I5 = I4 Then GoTo 105
For I6 = 1 To 10 Step 1
If I6 = I1 Or I6 = I2 Or I6 = I3 Or I6 = I4 Or I6 = I5 Then GoTo 104
For I7 = 1 To 10 Step 1
If I7 = I1 Or I7 = I2 Or I7 = I3 Or I7 = I4 Or I7 = I5 Or I7 = I6 Then GoTo 103
For I8 = 1 To 10 Step 1
If I8 = I1 Or I8 = I2 Or I8 = I3 Or I8 = I4 Or I8 = I5 Or I8 = I6 Or I8 = I7 Then GoTo 102
For I9 = 1 To 10 Step 1
If I9 = I1 Or I9 = I2 Or I9 = I3 Or I9 = I4 Or I9 = I5 Or I9 = I6 Or I9 = I7 Or I9 = I8 Then GoT
o 101
Let I10 = 55 - (I1 + I2 + I3 + I4 + I5 + I6 + I7 + I8 + I9)
T(I1) = "" + L(I1) + " " + L(I2) + " " + L(I3) + " " + L(I4) + " " + L(I5) + " " + L(I6) + " " +
 L(I7) + " " + L(I8) + " " + L(I9) + " " + L(I10) + " "

If I < S(1) Then GoTo 9
If I > (S(1) + 35) Then GoTo 7

Print I; L(I1); " "; L(I2); " "; L(I3); " "; L(I4); " "; L(I5); " "; L(I6); L(I7); " "; L(I8); "
 "; L(I9); " "; L(I10)

7 Write #1, I, T(I1)

9 If I = S(2) Then GoTo 4

I = I + 1
101 Next I9
102 Next I8
103 Next I7
104 Next I6
105 Next I5
106 Next I4
107 Next I3
108 Next I2
    Next I1
4 Close #1
Print "  --------- DONE ----------                * Use a MSWord or WPerfect to analyze and prin
t the total output, find the file at C:\INFER\INFER.TXT * "
End Sub

Private Sub cmdExit_Click()
Unload Me
End
End Sub

Private Sub cmdLIMITS_Click()
Cls
Prompt$ = "Enter Initial and Final bounds"
For J% = 1 To 2
    Title$ = "Boundary " & J%
    S(J%) = InputBox(Prompt$, Title$)
Next J%
End Sub

Private Sub cmdTextIN_Click()
```

Module1 - 1

```
Option Base 1
Public L(10) As String
Public T(3628800) As String
Public S(2) As Integer
```

Form1 - 2

```
Cls
Prompt$ = "Enter text."
For I% = 1 To 10 Step 1
    Title$ = "Text " & ((10 - I%) + 1)
    L(I%) = InputBox(Prompt$, Title$)
Next I%
End Sub
```

SOFTWARE

MODEL

Liberty Basic

Request free software, works with XP or WIN 7, at
gueller@aol.com

```
10 cls

print

print "Program MODEL.bas"

print "From POLYN: a1*x^8 + a2*x^7 + a3*x^6 + a4*x^5 + a5*x^4 + a6*x^3 + a7*x^2 +
a8*x^1 = C "

print : print

Print "Enter a1 = "

input a1

Print "Enter a2 = "

input a2

print "Enter a3 = "

input a3

print "Enter a4 = "

input a4

print "Enter a5 = "

input a5

print "Enter a6 = "

input a6

print "Enter a7 = "

input a7

print "Enter a8 = "

input a8

print "Enter C = "

input C

20 print "Enter value of x (1, 2, 3, or 4)"

input x

If x = 1 then goto 30

If x = 2 then goto 130

If x = 3 then goto 230
```

```
If x = 4 then goto 260

print : print

30 print "------------------------------- Case of x = 1): "

AA = a1

BB = a2

CC = a3

DD = a4

EE = a5

FF = a6

GG = a7

HH = a8

print

print " The a's are the same data: "

print

print AA,BB,CC,DD,EE,FF,GG,HH

print : print

40 Print " Another Run or Print ?"

print "Press 1 for Another Run, Press 2 to Print, or Press 3 to QUIT"

input a

If a = 1 then goto 10

If a = 2 then goto 320

If a = 3 then goto 400

print : print

130 print "------------------------------- Case of x = 2): "

AA = a1 / 256

BB = a2 / 128

CC = a3 / 64

DD = a4 / 32

EE = a5 / 16
```

```
FF = a6 / 8

GG = a7 / 4

HH = a8 / 2

print

print " The a's now are: "

print

print AA,BB,CC,DD,EE,FF,GG,HH

print : print

140 Print " Another Run or Print ?"

print "Press 1 for Another Run, or Press 2 to QUIT"

input a

If a = 1 then goto 10

If a = 2 then goto 320

If a = 3 then goto 400

print : print

230 print "-------------------------------- Case of x = 3): "

AA = a1 / 6561

BB = a2 / 2187

CC = a3 / 729

DD = a4 / 243

EE = a5 / 81

FF = a6 / 27

GG = a7 / 9

HH = a8 / 3

print

print " The a's now are: "

print

print AA,BB,CC,DD,EE,FF,GG,HH

print : print
```

240 Print " Another Run or Print ?"

print "Press 1 for Another Run, Press 2 to Print, or Press 3 to QUIT"

input a

If a = 1 then goto 10

If a = 2 then goto 320

If a = 3 then goto 400

print : print

260 print "----------------------------- Case of x = 4): "

AA = a1 / 65536

BB = a2 / 16384

CC = a3 / 4096

DD = a4 / 1024

EE = a5 / 256

FF = a6 / 64

GG = a7 / 16

HH = a8 / 4

print

print " The a's now are: "

print

print AA,BB,CC,DD,EE,FF,GG,HH

print : print

print "---"

300 Print " Another Run or Print ?"

print "Press 1 for Another Run, Press 2 to Print, or Press 3 to QUIT"

input a

If a = 1 then goto 10

If a = 2 then goto 320

If a = 3 then goto 400

320 lprint : lprint : lprint

```
322 lprint " Reduced values of a's for the new increased value of x = _____ "

324 lprint

326 lprint AA,BB,CC,DD,EE,FF,GG,HH

327 lprint

328 lprint " The values of a's for x = 1 were data. "

330 print "Press 1 for Another Run, or Press 2 to QUIT"

If a = 1 then goto 10

If a = 2 then goto 400

400 cls

print

dump

cls

end
```

About the Author

Sam Gueller was born in Argentina where he studied Land Surveying and Civil Engineering, later he continue studies at the Universities of Delft, Technion, Graz, Cincinnati and California. He has twelve years experience as Assistant Professor, worked for Consulting firms, and has been staff International Specialist in the Inter-American Development Bank.

He has working experience in fifteen countries, and interest in science and technology. He made math and analog modeling of several phenomena particularly in Fluid Mechanics and authored six books and presented papers in several Congresses, domestic and abroad. He is Life Member of ASCE, Member of the Asociación Física Argentina (AFA) and other professional societies.

Dr. Gueller and wife reside part time in USA or Argentina.

gueller@aol.com

www.ingramcontent.com/pod-product-compliance
Lightning Source LLC
Chambersburg PA
CBHW050718180526
45159CB00003B/1060